Studies in Applied Philosophy, Epistemology and Rational Ethics

Volume 80

Studies in Applied Philosophy, Epistemology and Rational Ethics (SAPERE) publishes new developments and advances in all the fields of philosophy, epistemology, and ethics, bringing them together with a cluster of scientific disciplines and technological outcomes: ranging from computer science to life sciences, from economics, law, and education to engineering, logic, and mathematics, from medicine to physics, human sciences, and politics. The series aims at covering all the challenging philosophical and ethical themes of contemporary society, making them appropriately applicable to contemporary theoretical and practical problems, impasses, controversies, and conflicts.

Book proposals can be sent to the Series Editor, Prof. Lorenzo Magnani at lmagnani@unipv.it. Please include: A short synopsis of the book or the introduction chapter A Table of Contents The CV of the lead author/editor(s) For more information, please contact the Editor-in-Chief at lmagnani@unipv.it.

Our scientific and technological era has offered "new" topics to all areas of philosophy and ethics – for instance concerning scientific rationality, creativity, human and artificial intelligence, social and folk epistemology, ordinary reasoning, cognitive niches and cultural evolution, ecological crisis, ecologically situated rationality, consciousness, freedom and responsibility, human identity and uniqueness, cooperation, altruism, intersubjectivity and empathy, spirituality, violence. The impact of such topics has been mainly undermined by contemporary cultural settings, whereas they should increase the demand of interdisciplinary applied knowledge and fresh and original understanding. In turn, traditional philosophical and ethical themes have been profoundly affected and transformed as well: they should be further examined as embedded and applied within their scientific and technological environments so to update their received and often old-fashioned disciplinary treatment and appeal. Applying philosophy individuates therefore a new research commitment for the 21st century, focused on the main problems of recent methodological, logical, epistemological, and cognitive aspects of modeling activities employed both in intellectual and scientific discovery, and in technological innovation, including the computational tools intertwined with such practices, to understand them in a wide and integrated perspective. **Studies in Applied Philosophy, Epistemology and Rational Ethics** means to demonstrate the contemporary practical relevance of this novel philosophical approach and thus to provide a home for monographs, lecture notes, selected contributions from specialized conferences and workshops as well as selected Ph.D. theses. The series welcomes contributions from philosophers as well as from scientists, engineers, and intellectuals interested in showing how applying philosophy can increase knowledge about our current world. Indexed by SCOPUS, zbMATH, SCImago, DBLP. All books published in the series are submitted for consideration in Web of Science.

Sanderson Molick
Editor

Demarcating Logic and Science

Exploring New Frontiers

Springer

Editor
Sanderson Molick
Division of Humanities
Federal Institute of Pará
Pará, Brazil

ISSN 2192-6255 ISSN 2192-6263 (electronic)
Studies in Applied Philosophy, Epistemology and Rational Ethics
ISBN 978-3-032-25673-7 ISBN 978-3-032-25674-4 (eBook)
https://doi.org/10.1007/978-3-032-25674-4

This Springer imprint is published by the registered company Springer Nature Switzerland AG
The registered company address is: Gewerbestrasse 11, 6330 Cham, Switzerland

Acknowledgements

The editor is grateful to many colleagues who contributed directly or indirectly to this project, including João Daniel Dantas, David Gomes, Sofia Abelha, Leon Commandeur, Kherian Gracher, Mansooreth Kimiagari, Hudson Benevides, Guilherme Araújo, Raoni Arroyo, and Edson Bezerra. Special thanks are also due to all the authors who contributed to this volume.

Introduction

Demarcation problems have long occupied a central place in philosophy. From Antiquity through the Modern era, philosophers have focused on the task of properly defining the boundaries between distinct domains of philosophical inquiry. In the twentieth century, these questions came to define debates in epistemology and the philosophy of science, especially in attempts to draw a principled line between science and non-science. However, when applied to logic, the demarcation problem assumes a distinctive form: is logic a part of science, a precondition for it, or something entirely distinct? Should we treat logical theories as fixed, a priori frameworks for evaluating reasoning, or can they be revised, assessed, and even rejected in the same way scientific theories are?

These questions have animated contemporary philosophy of logic through *antiexceptionalism about logic*. Originally defended by Timothy Williamson and Ole Hjortland, this philosophical position asserts that logical and scientific theories share a common methodology. They are both revised on the same grounds and methods. Historically, this view has deep roots, ranging from the mathematical empiricism of the eighteenth century to Russell's investigations into the role of inductive reasoning in logical theorizing, and even to Husserl's *Logical Investigations* and the broader debate concerning the normative role of logic within science.

In its contemporary form, the central historical root of logical anti-exceptionalism is found in Quine's work and the view that logic is not immune to revision, nor is it fundamentally distinct from the sciences in its methodology. Currently, the antiexceptionalist turn has led to a renaissance in the philosophy of logic, raising new questions about logical pluralism, theory choice, inferential norms, and the nature of logical disagreement. How can we choose among competing logical frameworks? What justifies adopting one logic over another? Can logical theories be empirically informed? These questions lie at the heart of the present volume.

Demarcating Logic and Science: Exploring New Frontiers brings together original contributions from philosophers working at the intersection of logic, epistemology, and philosophy of science. The chapters explore a wide range of perspectives on the boundary between logic and science, focusing on the implications of conceiving logic as a dynamic and revisable discipline.

The volume opens with César Frederico dos Santos, who explores a pragmatic view of logic as a cognitive tool, proposing that different logics serve different reasoning goals, and that their value lies in their use to perform certain reasoning tasks rather than in a metaphysical status. According to the author, viewing logics as cognitive tools is sufficient for overcoming the challenges faced by either the conventionalist or the representationalist views of logic. Elke Brendel then examines the epistemic virtues employed by logical abductivism when comparing rival logics. The liar paradox is discussed as a central desiderata for analyzing logical disagreements.

Gillman Payette and Nicole Wyatt contribute a chapter on bootstrapping logical theories, examining different strategies for avoiding the *background logic problem.* Jonas R. Becker Arenhart approaches the anti-exceptionalist view of *logic as models*, investigating how logical correctness arise from within logical practice itself. The chapter further considers alternative ways of framing data in scientific contexts.

Arenhart is also joined by Raoni Arroyo and Ederson Safra Melo in a further chapter disentangling scientific realism from logical anti-exceptionalism, clarifying the metaphysical commitments of each position. Marcos Silva's chapter delves into how reflective equilibrium and normative pressures shape logic revision. Finally, Ulf Hlobil closes the volume proposing a novel challenge to the anti-exceptionalist view by arguing that logic should not be regarded as a science, but as a discipline more closely aligned with philosophy and mathematics than with the empirical sciences.

Together, these contributions offer fresh perspectives on the question of how logic relates to science. We hope this volume will contribute to an emerging landscape in logical epistemology.

Pará, Brazil Sanderson Molick

Contents

Contributors

Raoni Arroyo Graduate Program in Philosophy, Universidade Federal de Santa Catarina (UFSC), Florianópolis, Brazil

Jonas R. Becker Arenhart Department of Philosophy, Federal University of Santa Catarina, Florianópolis, Brazil

Elke Brendel Institut für Philosophie, Universität Bonn, Bonn, Germany

César Frederico dos Santos Department of Philosophy, Federal University of Maranhão, São Luís, Brazil

Ulf Hlobil Department of Philosophy, Concordia University, Montreal, Canada

Ederson Safra Melo Research Group on Philosophy of Logic and Language (GFILL/CNPq), São Luís, Brazil;
Department of Philosophy, Universidade Federal do Maranhão (UFMA), São Luís, Brazil

Gillman Payette Department of Philosophy, University of Calgary, Calgary, Alberta, Canada

Marcos Silva Department of Philosophy, Federal University of Pernambuco, Recife, Pernambuco, Brazil

Nicole Wyatt Department of Philosophy, University of Calgary, Calgary, Alberta, Canada

Logics as Cognitive Tools

César Frederico dos Santos

Abstract Logical theories are usually seen as true or false in relation to the phenomenon they aim to describe, namely, validity. An alternative view suggests that the laws governing validity are "legislated-true," making logical theories conventional. In this chapter, I propose an intermediate standpoint in which logics exhibit a blend of descriptive and conventional aspects, balanced to fulfill their primary purpose, which is not seen as theoretical, but practical: assisting us in generating and identifying valid inferences. From this perspective, logics are cognitive tools, human creations designed primarily to facilitate the performance of cognitive operations. This viewpoint aligns logic more closely with technology than science and supports a pluralistic approach: the diversity of alternative technical solutions for issues related to the creation and recognition of valid inferences allows for the coexistence of multiple logics.

1 Introduction

The question of whether logical theories are true and, if so, true of what, is one of the most fundamental questions in the philosophy of logic. It is central to the debate monism *versus* pluralism, for example, since the existence of one or many true logics depends on whether a logic can be considered true in some sense. Answers to this question in the literature are varied, but most of them can be placed into either the representationalist or anti-representationalist camps. Representationalism, in the sense I am using this term here, is the view according to which logics are descriptive theories aimed at capturing traits of a preexisting reality. Frege (1960, 1977b), famously associated with Platonism, is among the best known advocates of representationalism about logic. The opposing view, which denies that logic aims to describe any existing reality, is best exemplified by Carnap (2001) and his conventionalism. According to him, logics are creative systems invented by logicians to serve practical purposes.

C. F. dos Santos (✉)
Department of Philosophy, Federal University of Maranhão, São Luís, Brazil
e-mail: cesar.frederico@ufma.br

S. Molick (ed.), *Demarcating Logic and Science*, Studies in Applied Philosophy, Epistemology and Rational Ethics 80, https://doi.org/10.1007/978-3-032-25674-4_1

Both representationalism and Carnapian conventionalism—the version of anti-representationalism I will consider here—face serious challenges. Representationalism presupposes the existence of a phenomenon targeted by logical theories and access to data about this phenomenon, but there is no consensus on the nature of such a phenomenon and what qualifies as relevant data for logical theorizing. Conventionalism, in turn, presupposes complete freedom to dictate what is considered true or valid in logic, but this is not substantiated in practice: certain choices are simply unacceptable.

In this chapter, I propose an alternative account of truth in logic, which combines aspects of both representationalism and conventionalism to address the shortcomings associated with their pure versions. I propose that logics are *cognitive tools*.

Tools are human inventions designed primarily to facilitate or enable the performance of certain tasks. Tools are not aimed at representing something, nor are they purely conventional. Instead, they need to combine representational and conventional aspects to achieve their purposes. Consider, for example, a typical hand tool, the screwdriver. If its tip does not fit the head of the screw, it fails to serve its practical purpose. The match between the screwdriver's tip and the screw's head can be viewed as a representational aspect of this hand tool. Other elements, such as the color of its handle, are conventional features. I submit that logics function as cognitive tools akin to hand tools in this regard: they are neither fully representational nor fully conventional, but rather exhibit a blend of representational and conventional aspects, weighted to enable the achievement of their purpose—namely, to help us overcome cognitive limitations in making inferences.

This chapter is structured as follows. In the next two sections, I discuss logical representationalism and logical conventionalism and their shortcomings. In Sect. 4, I characterize tools in general—using hand tools as paradigmatic exemplars—and discuss their representational and conventional aspects. In Sect. 5, I introduce the concept of cognitive tool, by analogy with the concept of hand tool. In Sect. 6, I explain how logics can be viewed as cognitive tools. In Sects. 7 and 8, I explore the representational and conventional aspects of logics, respectively, and discuss how these aspects matter to our understanding of truth in logic. As will be clear by the end of the chapter, the position defended here is a version of logical instrumentalism. In Sect. 9, I clarify my version of logical instrumentalism by comparing it with other versions found in the literature. Section 10 concludes the chapter with final remarks.

2 Representationalism

The gist of representationalism as an account of truth in logic is the idea that logical theories are intended to capture facts about a preexisting reality. The nature of such reality varies among authors: it may be viewed as a realm of mind-independent relationships between propositions (Priest, 2006), the formal structure of the world (Sher, 2016), inferential aspects of natural languages (Shapiro, 2006), or inferential aspects of mathematical practice (Martin and Hjortland, 2020). While the latter two are not human-independent realities, the crucial point is that they are conceived of as independent of the logician's activity. That is, even if inferential practices in natural languages and mathematics are somehow dependent on human factors, the task of the logician is viewed as confined to describing the logic "underlying" these practices. In this sense, these perspectives qualify as representationalist, as the term is employed here.

Representationalism has an important consequence for logical practice, often overlooked by its advocates. If logical theorizing is fundamentally a descriptive activity, then logicians engaged in this practice need data from the intended reality to inform their theories. In fact, this need is common to all theoretical activities whose purpose is descriptive, such as theorizing in the natural sciences. Accordingly, anti-exceptionalists (e.g., Hjortland, 2017) claim that there are no relevant methodological differences between theorizing in logic and theorizing in other sciences. In particular, they claim that theory choice in logic relies on abduction, much like theory choice in empirical sciences. The abductive method consists of selecting the best theory, i.e., the one that most effectively explains the available data. But what constitutes this data?

A major difficulty in addressing this question is that, in logical practice, we do not observe logicians employing data in a manner resembling the use of data in empirical sciences. Logicians do not systematically collect data, nor do they analyze data spreadsheets to formulate their theories. This fact alone casts doubt on both representationalism and anti-exceptionalism.[1] If logical theories are supposed to capture a phenomenon, as representationalism posits, how can logicians describe it without systematically collecting and analyzing data, as is done in other descriptive sciences?

A possibility is that the nature of the data considered in logical theorizing is notably different and unique compared to other sciences. In other words, while logic may be unexceptional methodologically, it could be exceptional in terms of the nature of its data basis. This raises the question of what these exceptional, unique data could be. Many candidates have been proposed, but here I will consider, as examples, only two candidates advanced by Martin and Hjortland (2023, p. 12): paradoxes and linguistic data. Let us consider each in turn.

First of all, it must be noticed that, if paradoxes such as the Liar, Sorites, and Russell's paradox do count as data, they constitute an exceptional form of data compared to data used in other sciences (e.g., they are not amenable to statistical treatment). Yet, Martin and Hjortland (2023, p. 12) claim that "[t]here is general acceptance within the logical literature that these paradoxes constitute data which logics must accommodate." Although it is undeniable that paradoxes play a central role in logical theorizing, their classification as data is not indisputable.

The hallmark of data is that it possesses, or is perceived to possess, a certain degree of independence from the theories that describe them. Paradoxes, however,

---

[1] It must be said that, by now, the term "anti-exceptionalism" has been extensively discussed in the literature and gained many different definitions. Here, I'm discussing what has been called "methodological anti-exceptionalism".

only emerge when a logical theory (even if inchoate) is in place. One could argue that paradoxes are theory-laden data, observable only through the lens of a logical theory. After all, even in the empirical sciences data has been considered to be theory-laden. Nevertheless, certain paradoxes are so theory-dependent that they are more accurately described as theoretical *artifacts*, if one insists on categorizing them as a form of observation at all. I submit that, rather than being any kind of observation, paradoxes are better viewed as theoretical shortcomings that must be addressed by improved theories. Seemingly, this was Russell's attitude towards the paradox named after him in Frege's system. Russell did not interpret the presence of the paradox as an indication of a fact of the matter, a datum about arithmetic, but rather as a flaw in Frege's theory that needed correction. Dialetheists aside, paradoxes are more likely to be intrinsic to the theory than representative of the phenomenon.

As for linguistic data, Martin and Hjortland (2023, p. 12) note that "[l]ogicians who seek to formalize natural language expressions, such as 'if ... then ...' and 'not', rely on competent speakers' linguistic judgments as empirical evidence." In contrast to paradoxes, linguistic data is not exceptional when compared to data used in other sciences. It is empirical and can be systematically collected using techniques such as corpus analysis. However, logicians typically do not systematically collect linguistic data or analyze it statistically. Most logicians, when appealing to evidence about natural language usage, often rely on their own linguistic intuitions. It is as if they were relying on a single data point to inform their theories. If the aim is to provide an accurate description of inference practices in natural languages, this reliance on a single data point becomes a serious limitation for any data-informed investigation.

In linguistics, scholars have faced criticism for relying too heavily on their own linguistic intuitions instead of conducting rigorous data collection (Maddirala, 2014; Stokhof and van Lambalgen, 2011). While this is a serious methodological concern for linguists, whose goal is to describe the practices of speakers, logicians can counter such criticism by arguing that logical theories do not seek to describe how natural languages are used but rather prescribe how they ought to be used for making valid inferences. In other words, they may argue that logical theories are normative rather than descriptive. In this case, competent speakers' linguistic judgments might not hold much significance. In fact, the limited importance logicians assign to the systematic treatment of linguistic data suggests that the discipline's purpose is not to provide a faithful account of language usage. If this is so, linguistic data may not be central to logic.

A way of reconciling the claim that logic is normative with regard to inferential practices and yet descriptive is to postulate the existence of a realm of mind-independent validity relationships between propositions. However, if propositions are thought of as abstract entities existing outside of space and time, accessing data about them becomes even more problematic. How can spatiotemporal human beings have knowledge of non-spatiotemporal propositions? Benacerraf's Problem awaits those who choose this route (Benacerraf, 1983).

Certainly, philosophers have suggested many other potential candidates as data in logic, which space prevents me from addressing here. However, I believe that the very fact that there is an intense debate on what constitutes data in logic—where a variety of proposals has been advanced, many of them incompatible with each other, and there is no sign of consensus in the near future—suggests that data plays a minimal role in logic. If data were really important for logical theorizing, so many philosophers would not miss this point in this manner. Furthermore, there are deep disagreements among scholars about the target phenomenon of logic and, consequently, about what counts as the relevant data. As (Hlobil, 2021) puts it, which data one deems relevant depends on one's conception of logic, and this in turn depends on one's views regarding vexed disputes about the nature and aims of logic.

In summary, representationalism, as an account of truth in logic, presupposes the existence of a phenomenon (or multiple phenomena) targeted by logical theories and access to data about this phenomenon. However, there is no universally accepted opinion about the nature of this phenomenon and what counts as relevant data for logical theorizing. The strongest candidates as data for logic either do not properly qualify as data (the paradoxes) or do not appear to be relevant for logical practice (linguistic data). These challenges cast doubt on representationalism as a correct account of truth in logic. Let us now turn our attention to conventionalism.

3 Conventionalism

The gist of conventionalism as an account of truth in logic is the idea that logicians are free to legislate what counts as logically true. Logical principles are seen as "true by convention," with no commitment to describing a preexisting reality. One prominent advocate of this view is Carnap. His well-known Principle of Tolerance states that

> *In logic, there are no morals*. Everyone is at liberty to build up his own logic, i.e., his own form of language, as he wishes. All that is required of him is that, if he wishes to discuss it, he must state his methods clearly, and give syntactical rules instead of philosophical arguments. (Carnap, 2001, p. 52)

That is, when proposing a logic, there is no need to present arguments to the effect that the logic correctly captures a supposed underlying phenomenon or any other philosophical arguments. Logic is not representational and, as a result, there is no right or wrong choice in logic.

One of the main objections to this sort of extreme conventionalism is the observation that, in practice, there is no total freedom to legislate what qualifies as logically true. Prior's (1960) tonk connective is the most famous counter-example to the idea of absolute freedom in logic. The use of tonk is governed by two rules: (a) from P one may infer P tonk Q; and (b) from P tonk Q one may infer Q. Together, these rules imply that from any P, one may infer an arbitrary Q, under all and any circumstances. Although the syntactical rules governing tonk are clearly stated (fulfilling Carnap's criterion for applying the principle of tolerance), almost no one is prepared to accept tonk as a legitimate logical operator.

Warren (2020) presents an alternative conventionalist approach aimed at addressing the shortcomings of previous forms of logical conventionalisms, such as Carnap's. However, his conventionalism differs from Carnap's on a major point. Warren defines logical conventionalism as the thesis that "the facts about logical truth, logical falsity, and logical validity in any given language are fully explained by the linguistic conventions of that language," and adds: "As I understand it, logical conventionalism is a thesis about logical truth and validity in natural languages" (Warren, 2020, part I, ch. 1, II). As these quotes indicate, Warren believes that the primary locus of convention lies within natural language, rather than within the work of logicians, as Carnap believed. Warren assumes that natural languages have a underlying logic, established by convention, and that the task of the logician is to identify these conventions already present in natural language. As he puts it, "the formal methods involving proof construction *model* our actual processes of inference and argument construction" (part I, ch. 2, IV; emphasis added). In this way, Warren denies the logician the Carnapian freedom to devise her own logic; rather, logician's work is confined to describing conventions established within natural language. While Warren acknowledges that logical principles ultimately stem from conventions, his conception of the logician's role is representationalist.

Be that as it may, Warren offers an ingenious philosophical argument showing that a natural language incorporating tonk is not as objectionable as commonly perceived. In doing so, however, he concedes that such a language would be "a poor language …practically useless for most imaginable purposes" (Warren, 2015, p. 22). In my view, this is the main challenge to conventionalism: what stands in the way of the freedom to legislate what counts as logically true, both in natural and artificial languages, are pragmatic concerns. Although the freedom to specify syntactical rules may be complete, certain rules are simply "poor" or "useless," even if the conventionalist refrains from describing them as plainly false. Carnap himself acknowledges that "[t]he acceptance or rejection of abstract linguistic forms …will finally be decided by their efficiency as instruments" (Carnap, 1983, pp. 256–7).

Judgments of efficiency and effectiveness are always relative to a goal. Conceding to conventionalism, it is not difficult to grant that logicians are free to stipulate their goals, at least within a broad range of options. As (Dutilh Novaes, 2021, p. 83) puts it, "[w]hat defines which logic is the 'right one' are the motivations of participants when engaging in a given dialogical situation, and their mutual agreement in terms of the structural and logical features of that particular conversation." However, this cannot be the whole story. If there are better or worse choices given a goal, then there must be other, possibly non-arbitrary, conditions constraining the efficiency and effectiveness of these choices. There are objective facts that explain why some choices are better than others in view of a given objective. Regarding logical systems, Carnap hints at this when he says that logical systems must be judged "by their efficiency *as instruments*," as quoted above. In the next section, I examine the objective factors contributing to judging the efficiency and effectiveness of instruments in general.

4 Tools

Hand tools, the most paradigmatic examples of tools, are human creations whose primary function is to enable or facilitate the performance of manual operations. In this section, an analysis of hand tools will help illustrate the key properties of instruments and tools in general. In the following sections, we will see that these properties are also inherent in cognitive tools and, particularly, in logic.

The most distinctive trait of tools, the property that sets them apart from other things, is that they are designed for a specific purpose (Schmitt, 1965). However, fulfilling a purpose is not the sole reason why we create tools. If we could achieve the intended purpose using only our bodily means, a tool would be dispensable. Consider a shovel, for instance. A shovel facilitates or enables the performance of manual operations for which our bare hands and arms are impractical or limited. While our bare hands can dig in soft soils, such as beach sand, they are inadequate for digging in hard soils. They are too fragile, and our arms are too weak for such a task. By using a shovel, however, we can overcome these limitations. The shovel's blade allows us to cut into hard soils, and the handle amplifies the force of our arms, making it easier to lift and toss the shovel's content. The shovel adds to our bodily skills features that we lack: hardness, sharpness and incremented force. In general terms, hand tools enhance, extend or add new features to our bare hands. They are created in response to a limitation in the preexisting means (our hands, in the limiting case), and they help overcome such limitation.

In the customary distinction between inventions and discoveries, the former are often seen as involving a high degree of creative freedom—conventionality—whereas the latter are regarded as the mere revelation of an existing, independent reality. Tools fall into the category of inventions and, as such, many of their aspects are conventional. However, tools are not freely created—they differ, for example, from pieces of abstract art, the ultimate example of artifacts created with total freedom. Tools have a practical purpose to fulfill; they must be functional. Attaining functionality imposes constraints on their design. The design of a tool is subject to what Kant described as hypothetical or technical imperatives (Kant, 1998, 4:415–16). Technical imperatives are if-then clauses in which the antecedent states a goal and the consequent states a means to achieve the goal. For example: "if you want to dig in hard soil, you need an instrument with a sharp and hard blade." This conditional statement conveys a command with a biding effect on those seeking to achieve the goal. Someone designing a tool for excavating in hard soil is not entirely free to invent but is constrained by this technical imperative.

The link between antecedent and consequent in a technical imperative arises from practical considerations concerning both the environment and the agent. The determination of suitable means to achieve a practical end is not dictated by free conventions but rather by the capabilities of the agent pursuing that end and the relevant properties of the environment where the end is to be realized. The shovel, for example, must not only feature a sharp and hard blade—due to soil conditions—but also a suitable handle—due to user's conditions. A shovel with a handle causing significant discomfort to the user is not functional. In essence, a tool must "fit," so to speak, the environment where it is intended to be used, as well as suit its users. Tool design may thus be constrained by various technical imperatives related to environmental fitness and user suitability.

This does not imply, however, that there is no room for conventions in tool design. Up to this point, I have loosely associated conventionality with complete creative freedom, following a literal interpretation of Carnap's Principle of Tolerance. We have seen that there is no complete creative freedom in tool design. However, even when complete creative freedom is absent, the existence of alternatives makes room for conventions. Accordingly, Azzouni (2014, p. 36) defines a practice as conventional when it follows one course of action selected among others, different but roughly equivalent courses, concerning their efficiency and effectiveness. In other words, if there are multiple ways to achieve a certain goal, and they are roughly equivalent in terms of efficiency and effectiveness, the choice of following one over the others is up to personal preferences and is therefore conventional. Azzouni provides an example of a tribe using the technique of rubbing two sticks together to start a fire. If this tribe lives in an area with available rocks, so that they could start a fire by banging rocks together, their habit of using sticks becomes a matter of convention. However, if the tribe lives in an area without rocks, banging rocks together is not a viable alternative technique, and therefore the practice of rubbing sticks is not conventional for them. In situations where there is obligation or necessity, where alternative courses of action are absent, there is no convention. In such a scenario, the practice is entirely governed by technical imperatives. In contrast, when a range of alternatives is available, even if limited, and even if technical constraints apply, there is room for conventions.

Once conventionality is viewed as the existence of alternatives, it is easy to see that the conventional aspects of hand tools are not limited to decorative features, such as color and notches. Consider, for example, the task of fastening a nut. One may opt between using a ring spanner (wrench), an open-ended spanner, or an adjustable spanner. Insofar as all of them are equally suitable for the nut, the choice is conventional. This does not mean, though, that the choice is arbitrary. Personal preferences, secondary objectives, tradition, and possibly other factors can explain it. Secondary objectives are especially important, since they can turn what was originally conventional into mandatory. For example, if the purpose was not only to fasten a specific nut, but also to acquire a versatile tool that can be used to fasten nuts of different sizes, then choosing the adjustable spanner is mandatory. Conventionality is not absolute, but relative to one's objectives.

One's objectives along with characteristics of the environment and the agent impose technical imperatives or technical constraints, which limit the range of alternatives. The alignment that there must be between a tool, its user and the usage environment gives rise to the presence of representational aspects in tools. Screwdrivers exemplify this point nicely. The tip of the screwdriver must fit the head of the screw. If it does not, the screwdriver cannot function as intended. Similarly, the handle of the screwdriver must conform to the void inside a closed human hand, making it easy to grasp. From the tip of the screwdriver and its handle, one can infer aspects of the screw head (whether it is Philips, slotted, hexagonal, etc.) and aspects of a typical human hand (its size, for example). In a sense, the tip of the screwdriver *represents* aspects of screw heads and its handle *represents* aspects of human hands.

One may question the true representational nature of these aspects. To clarify what I describe as the representational aspects of tools, I adopt Menary's (2007) definition of representation, based on what he terms "the Peircean principle." According to this principle, a representation is anything that "involves the following three components: a representational vehicle with representationally salient properties, an object or environmental property and a consumer that exploits the vehicle in virtue of its representationally salient properties" (Menary, 2007, pp. 95–96). This definition broadens the scope of what qualifies as a representation; essentially, anything perceived by an agent as standing for something else is considered a representation. Thus, not only signs in the strict sense, but also material objects as diverse as sculptures or the disposition of glasses on a table, fall into the category of representations. For example, if a bartender sees the disposition of glasses on the counter as representing the drinks consumers have recently ordered, then this disposition is a representation. In the case at hand, a screwdriver's tip and handle serve as representational vehicles insofar as a consumer perceives them as standing for a type of screw head and the shape of a human hand, respectively. This could occur, for instance, when someone is searching for a suitable screwdriver within a toolbox.

If this is so, we may say that tools display a combination of representational and conventional aspects, which are interconnected. The freedom to design a tool is limited by the representational aspects it must include. For instance, the permissible range of shapes the handle of a shovel may take is restricted by features of the human hands and arms, and shovels' handles represent aspects of the human hands and arms to the extent that they are adequate for use by human hands and arms. By the same token, the permissible range of shapes the tip of a screwdriver may take is determined by the requisite of adequacy to the shape of the head of the intended type of screw, and the screwdrivers' tip represents the head of certain screws to the degree that it is suitable for screwing them.

Restating the main points, we have seen that tools are human inventions that enhance, extend or add new features to our bodily skills. They are created in response to a limitation in the preexisting means (primarily our bodies), and they help overcome such limitation. They do so by blending conventional and representational aspects. The need of incorporating in the tool representations of properties of the user and properties of the usage environment to ensure its functionality limits the range of conventions available to an inventor when designing a tool. The specific properties that constrain the design depend on the purpose of the tool. The efficiency and effectiveness of the tool hinge on how it integrates conventional and representational aspects to ensure its functionality.

5 Cognitive Tools

As hand tools are human creations designed primarily to enable or facilitate the performance of *manual* operations, cognitive tools are human creations designed primarily to enable or facilitate the performance of *cognitive* operations. Analogous to hand tools, cognitive tools enhance, extend or add new features to our "brains" (as opposed to our hands) and they are created in response to limitations of our cognitive functions (rather than limitations of our bodily skills). A prototypical example of a cognitive tool is an abacus. Although we can perform arithmetical operations mentally, without the aid of external tools, an abacus enables its user to perform arithmetic operations faster and more accurately than they could by relying only on mental operations. Similar to a shovel that enables its user to dig in harder soils, an abacus empowers its user to calculate with larger, and therefore harder, numbers.

Despite these similarities, cognitive tools differ from hand tools in a key aspect: they do not need to be three-dimensional, handleable external objects. Cognitive tools can be *techniques*, i.e., rule-based methods of performance that facilitate the fulfillment of cognitive tasks. Typically, these techniques involve symbols manipulated either on external media (e.g., on paper) or mentally. Although techniques lack a three-dimensional external body, typical of other tools, what justifies describing them as "tools" is a perceived similarity to handleable external tools in that both are human inventions created to boost our capacities.[2]

One example of a cognitive tool that is not a handleable artifact is counting. It is a well-established result in numerical cognition that, by using only perception, we are unable to asses the exact cardinality of collections with more than three or four items (Knops, 2020). In order to overcome this limitation, prehistoric humans created tallying and counting techniques (dos Santos, 2021a; Overmann, 2023). The counting technique involves reciting the counting words in ascending order while a one-to-one correspondence is established between the words of the sequence and the items of the collection being counted. The last word recited represents the cardinality of the collection. This technique is a cognitive tool in its own right (dos Santos, 2025). Another example of a cognitive tool that is not a three-dimensional artifact is the notation system of Arabic digits combined with the algorithms for arithmetic operations. These algorithms exploit the properties of the decimal place-value system to permit accurate pencil-and-paper and mental calculations. Counting and the system of Arabic digits provide rule-based methods of performance—techniques, cognitive tools—that not only facilitate but indeed make exact calculation and the assessment of the exact cardinal size of collections possible. These techniques, just like an abacus or other handleable tools, enhance, extend, and add new features to the innate, genetically evolved power of our brains.

[2] For definitions of cognitive tools in accordance with the one adopted here, see, e.g., (Dascal, 2002), who defines cognitive technologies as "every systematic means—material or mental—created by humans that is significantly and routinely used for the performance of cognitive aims," and (Menary and Gillett, 2022).

Cognitive tools also combine representational and conventional aspects to ensure functionality. For example, whether we count the beads of a necklace clockwise or anti-clockwise is a matter of convention, while the requirement that each bead must be paired with a single counting word is a matter of representation. The representational requirement is that each word in the counting sequence must represent precisely one bead, ensuring that an initial segment of the sequence can model or represent the total number of beads in the necklace.[3] Logic provides another good example of a combination of representational and conventional aspects, as we will see below. Firstly, though, let me explain why logics qualify as cognitive tools.

6 Logics as Cognitive Tools

At least since Aristotle's works on logic became collectively known as the *Organon*, or tool of thought, the instrumental role of logic is widely recognized. Therefore, there should be nothing controversial in considering logic as a cognitive tool. In this section, I emphasize the similarities between logics and tools in general, enabling us to investigate the representational and conventional aspects of logic in the following two sections.

As discussed in Sect. 4, what distinguishes tools from other human creations is that they are designed for a specific purpose that is not easily achievable using only our bodily and cognitive skills. This characterization nicely applies to logic. Although the contemporary development of logic may be guided by multiple goals, traditionally logic is concerned with validity. The identification and production of arguments where it is impossible for the premises to be true while the conclusion is false have motivated the creation of logics since Ancient Greece. Unsurprisingly, a plethora of studies in experimental psychology during the last century has shown that the recognition of validity is not a task where humans are naturally proficient (Evans et al., 1993). Rather, when we are not assisted by logical tools, we are subject to the influence of several cognitive biases that hinder our capacity to distinguish valid from invalid arguments. The development of logic can be seen as responding to this cognitive limitation.

One of the most well-documented obstacles to the unassisted evaluation of validity is the so-called *belief bias*. It is described as "the tendency to automatically bring prior knowledge to bear when solving problems" (Stanovich, 2003, p. 292). While relying on prior knowledge is not a bias in most situations, it can be detrimental to the evaluation of validity. Validity is a relationship between the premises of an argument and its conclusion. Accordingly, the evaluation of validity must be confined to assessing whether the premises *alone* imply the conclusion. In other words, we have to "bracket" our previous beliefs and take the premises at face value, evaluating

[3] For a detailed description of the counting procedure and the role of the sequence of counting words as a model collection, see (dos Santos, 2023, sec. 2).

exclusively the connection between them and the conclusion, regardless of the plausibility of the argument in face of prior knowledge (Dutilh Novaes, 2021, pp. 7–9). However, putting prior beliefs aside to evaluate validity is not something people in general can easily do, due to belief bias. "[T]his tendency toward contextualizing problems with prior knowledge is so ubiquitous [in human cognition] that it cannot easily be turned off" (Stanovich, 2003, p. 292). Stanovich describes belief bias as "a fundamental computational bias (one that pervades virtually all thinking, whether we like it or not)" (Stanovich, 2003, pp. 292–3).

A simple illustration of how belief bias affects the evaluation of validity is demonstrated in a study conducted by Sá et al. (1999). They tested the ability to reason logically of individuals who had undergone normal schooling but had never received specific training in logic. Participants were presented with syllogistic arguments and then asked to evaluate their validity. The results showed that, rather than considering only the relationship between premises and conclusion, participants preferred to rely on their prior beliefs about the topic of the argument. For example, when asked to evaluate the following argument:

A)
All living things need water.
Roses need water.
Therefore, roses are living things.

68% of the participants judged the argument as correct, despite its invalidity. Since the conclusion did not challenge their prior beliefs, they simply accepted the argument as correct.

This does not mean that people not trained in logic are doomed to reason illogically. In another test conducted by Sá et al. (1999), participants performed better. This time, they were presented with an argument displaying the same logical form but containing made-up words:

B)
All animals of the hudon class are ferocious.
Wampets are ferocious.
Therefore, wampets are animals of the hudon class.

In this case, 78% of the participants noticed that the argument is invalid. In B, the familiar expressions "living things" and "roses" were replaced by the made-up words "hudon" and "wampets." Since participants did not know the meaning of these words, they could not rely on previous beliefs to evaluate B. Thus, they had to stick to the information contained in the premises only. By doing so, they noticed that the argument is invalid. In other words, these made-up words produced a debiasing effect by suppressing the influence of prior beliefs (Dutilh Novaes, 2012, p. 137).

Dutilh Novaes (2012, ch. 7) points out that formalisms produce a similar debiasing effect. For example, consider the following semi-formalization of syllogism *A*:

C)
All ϕs need water.
ψs need water.
Therefore, ψs are ϕs.

Insofar as ψ and ϕ are symbols denoting arbitrary, unknown classes, the effect will be the same as that obtained by using the terms "hudon" and "wampets" in B. These symbols do not evoke prior beliefs that could influence the judgment of the validity of the argument. This is a reason why we resort to formalisms when analyzing validity: they help us counter belief bias (Dutilh Novaes, 2012).

Certainly, individuals with more disciplined reasoning processes, trained in scientific, mathematical or philosophical investigation, are unlikely to fall prey to belief bias in arguments as simple as these. However, they may succumb to more complex and long chains of inference. Philosophers, scientists and mathematicians are as susceptible to reasoning biases as everyone else. In mathematics, in particular, the influence of belief bias can make the task of proving theorems, especially in the case of long proofs, very tricky. A sympathy for a tacit principle or the belief in an as-yet-unproved result may affect logical reasoning and lead to invalid proofs.

It is not an exaggeration to say that the first modern logic was created to counter belief bias. It is well-known that Frege created the *Begriffsschrift* in the course of his broader project of reducing arithmetic to logic. His goal was to establish a solid foundation for arithmetic by demonstrating that all of arithmetic could be proven solely from logical axioms. In attempting this, he faced two difficulties. First, he had to contend with the imprecision of natural language. Second, he had to address the challenge of preventing illogical leaps and the intrusion of spurious assumptions in long chains of inferences. The *Begriffsschrift* was designed to address these two needs.

> The course I took was first to seek to reduce the concept of ordering in a series to that of logical consequence, in order then to progress to the concept of number. So that nothing intuitive could intrude here unnoticed, everything had to depend on the chain of inference being free of gaps. In striving to fulfil this requirement in the strictest way, I found an obstacle in the inadequacy of language: however cumbersome the expressions that arose, the more complicated the relations became, the less the precision was attained that my purpose demanded. Out of this need came the idea of the present *Begriffsschrift*. It is thus intended to serve primarily to test in the most reliable way the validity of a chain of inference and to reveal every presupposition that tends to slip in unnoticed, so that its origin can be investigated. (Frege, 1997a, pp. 48–49)

His attempt to counter belief bias is clear. "Interestingly, and while he was obviously unaware of the concept of belief bias as such, one of Frege's main motivations to develop a special notation for scientific inquiry was precisely to counter our tendency to let presuppositions 'sneak in'" (Dutilh Novaes, 2012, p. 153).

If Frege's concerns are representative of the motivations of the logicians who founded the discipline, we may say that logics were created as cognitive tools designed to identify and produce valid inferences in face of our natural difficulty recognizing validity due to belief bias. Certainly, belief bias is not the only cognitive limitation logic helps us overcome. Frege himself mentioned the imprecision of natural language as another difficulty. Furthermore, experimental psychologists have identified various other error tendencies affecting deductive reasoning (Pohl, 2017). Formalisms, explicit logical principles, and deterministic rules of inference function as regulatory techniques that help us counteract such influences, making inferences as objective and free of personal influences as possible. In the next two sections, we will explore how logic achieves this by combining representation and conventionality.

7 Representational Aspects of Logics

In Sect. 4, we saw that, for a tool to be functional, it must align with its intended domain of application. This alignment between the tool and its usage environment constitutes what I earlier defined as the representational content of a tool. If logics are indeed cognitive tools, as argued above, it prompts us to inquire about their representational contents and what exactly they represent. This is what I investigate in this section.

Let me begin by clarifying the notion of *domain of application*, as I am using it here. In logic, this notion is usually understood in tandem with the distinction between pure and applied logic. A pure logic is defined as a structure comprising a formal language and a consequence relation established on sentences of this language. This structure is a purely formal, uninterpreted mechanism. Pure logics have no representational content. An applied logic, in turn, is defined as an interpretation of a formalism "in some way or other," as Priest puts it (Priest, 2006, p. 195). The domain of application of a logic is defined, here, as the interpretation assigned to a pure formalism. This interpretation need not reference an existing reality, nor involve the practical application of the logic. The sole requirement is that it interprets the formalism in terms relevant to argumentation; otherwise, it would not be appropriately classified as logic (Haack, 1978, p. 3).

The representational content of an applied logic comes from the aspects of its domain of application that it needs to incorporate to function properly. The primary function of deductive logic is the transmission of truth from the premises to the conclusion of arguments. How can this function be fulfilled? I draw on (Sher, 2016) (but not subscribing to her logical realism) to answer this question. The general idea is that for truth transmission to be ensured, the mechanisms of a logic must be "anchored" in regularities of its domain of application. In other words, the mechanisms of inference must *represent* certain regular properties of the domain. These regular properties are those "connecting the circumstances that have to hold in the world [domain of application] for the premises of a given logical consequence to be true to the circumstances that have to hold in the world [domain of application] for its conclusion to be true" (Sher, 2016, p. 271). While I do not endorse Sher's realism, I recognize that she gives a good explanation of the representational relationship between a logic and its domain of application. Where she posits the world as the rightful domain of application for logic, I leave this open: the domain of application of a logic depends on one's interests and goals. Presenting Sher's view in more metaphysically neutral terms, the basic idea can be explained as follows.

Let A and B denote certain circumstances in a given domain. If it is known that whenever A is the case, B is also the case, one can make use of this knowledge to build an inference mechanism in a formalism $\mathfrak{L}$. This is done by representing A and B in the language of $\mathfrak{L}$, obtaining $\mathcal{A}$ and $\mathcal{B}$, and including $\mathcal{A} \models \mathcal{B}$ among the laws governing the production of inferences in $\mathfrak{L}$. As yet, though, there is nothing specifically *logical* in such a mechanism of inference. The relationship between the circumstances A and B can be governed by mathematical, physical, or even political laws. For example, if the domain of application of $\mathfrak{L}$ is the set of natural numbers, this could be the relationship captured by $\mathcal{A} \models \mathcal{B}$:

1. whenever n is even (A), $n + 1$ is odd (B)

If the domain of application comprises only people, this could be the relationship captured:

1. whenever a is English (A), a is British (B).

Consequence relations based on such regularities can be called *materially* valid (Priest, 2023, p. 455). However, logics are usually concerned with *formally* valid consequence relations. This means that the types of regularities logics typically focus on are not the ones exemplified above; instead, logics are concerned with what Sher describes as *formal* regularities. She explains formality in terms of invariance. Roughly, formal properties are those that are invariant under one to one replacements of individuals, regardless of the type of individuals (Sher, 2016, p. 276 ff.). For example, "[i]f you have a formal structure of people and you replace the people, in a 1–1 manner, with, say, numbers, the formal structure will not change" (Sher, 2016, p. 276). Clearly, this does not apply to the aforementioned examples because the properties they exploit are not invariant across all types of individuals. Thus, a mechanism of inference based on such regularities could have a very limited scope of application (numbers and people, respectively).

The request for a wide scope of application—ultimately, for the widest possible scope—is what renders the exploitation of formal properties so valuable for logic. If we seek a tool to aid the production and recognition of valid inferences about no matter what, its mechanisms of inference must be anchored solely in properties shared by everything. But what are these properties? Sher gives several examples:

> In every area of the world individuals are *identical* to (*different* from) individuals; objects lie in the *intersection*, *complement*, and *union* of properties; properties are *included* in properties; relations are (or are not) *transitive*, *reflexive*, *well-ordered*; properties and relations are *non-empty* or *universal*, *finite* or *infinite*, hold of *n* elements (tuples of elements) or *m* elements, and so on. Given the prevalence of formal properties and relations, the possibility of developing a wholesale method for extending knowledge based on universal laws governing the formal properties of objects arises. Such a method will enable us to move from existent beliefs to new beliefs without loss of truth and regardless of field of interest. The task of logic, on this approach, is to construct a theory of the transmission (preservation) of truth based on formal or structural grounds. (Sher, 2016, p. 326)

Sher is a realist, and as such she regards these properties as structural properties of the world. Nevertheless, this is not essential for my point here. One could equally view them as structural properties of the linguistic framework we employ to organize our experiences, or as products of Kant's categories of understanding, or even as useful fictions. Be that as it may, my point is that, if one seeks to set up a general tool for making inferences—"a wholesale method for extending knowledge," as Sher describes it—she needs to exploit such structural properties to ground the mechanisms of inference of her formalism.

One might object that, if these properties are categories of understanding or mere useful fictions, a logic incorporating them fails to represent anything. Recall, though, that tools must not only suit their usage environment but also their users, as mentioned in Sect. 4. Thus, certain representational aspects they incorporate may come from the user—like the handle of a screwdriver, which represents the grip of a closed hand. While realists claim that the structural properties mentioned by Sher in the quotation above mirror reality, Kantians may argue that they reflect the categories of understanding. Anyway, for our purposes here, it suffices to note that these structural properties, regardless of their origin, are good examples of the representational aspects that logics must consider.

Sher (2016, p. 273) provides straightforward examples of how these properties can be exploited by logic. Let B and C be the extensions of any two properties in a certain domain of application. If it is known that, in this domain, the intersection of B and C is non-empty, we can conclude—given the structural properties of the domain—that B is also non-empty. This observation about the domain underlies the logical consequence $\exists x(Px \wedge Qx) \vDash \exists x Px$. Now that we know that B is non-empty, we can infer—given the structural properties of the domain—that the union of B with any other extension D will also be non-empty. This forms the basis for the logical consequence $\exists x Px \vDash \exists x(Px \vee Qx)$. Similar reasoning can explain instances of failed logical consequence. For example, a proposed law of logical consequence stating that one can infer $\neg Py$ from Px and $x \neq y$ fails to ensure truth transmission because it is not a fact (at least in most domains) that there is no property common to multiple objects (Sher, 2011, pp. 358–9).

Maddy (2007) offers further examples of structural properties a logic must take into account if its domain of application encompasses the entire world as we know it. Similar to Sher, Maddy posits that logical laws represent fundamental aspects of the structure of reality. According to Maddy, our world exhibits a "Kantian-Fregean (KF) structure." Worlds displaying this structure, which she terms KF-worlds, consist of

> objects that bear properties and stand in relations to one another; that some of these properties could be universal, holding of all objects; that some situations involving these objects would be complements (corresponding to the logical form of negation), conjoinings (corresponding to conjunction) or disjoinings (corresponding to disjunction) of others; and that some interrelations between these situations would be robust ground-consequent dependencies. This much fairly minimal logical structure is enough to validate many classical inferences involving negation, conjunction, disjunction and the quantifiers: introduction and elimination rules, distributive laws, and DeMorgan equivalences, for example. (Maddy, 2014, pp. 16–17)

Maddy's point is that the world itself is organized in "objects that bear properties and stand in relations to one another," and logic makes use of this organization to ground its laws of deductive inference. Once more, we do not need to subscribe to Maddy's realism. For the purposes at hand, it doesn't matter whether the world is inherently organized this way or if it's merely how we structure our experiences.

The key point is that, once the intended domain of application of a logical tool is thought of as structured in this manner, a logic can use these aspects in its inference mechanisms.

Additionaly, Maddy observes that the coupling between a logic and its domain of application is not always total. Sometimes, idealizations and simplifications are needed. For example, Maddy claims that the law of the excluded middle is not accurate as a description of the structure of the world, given the existence of fuzzy objects and properties, such as frogs and tadpoles (Maddy, 2007, p. 241). Since dealing with vague predicates can be complex, vagueness is idealized away in classical logic (Maddy, 2007, pp. 286–287), for the sake of simplicity. This is a deliberate design decision.

Another point Maddy highlights is that the macro-world, characterized by a KF structure, and the micro-world, governed by quantum mechanics, are significantly distinct from each other, therefore potentially necessitating different logics. A logic rooted in the structural properties of KF-worlds "cannot be assumed to apply effectively in every situation; it will only deliver validity when the relevant KF-structure is present" (Maddy, 2007, p. 271). Such a logic "will lead us astray when we attempt to speak of subatomic particles, their properties and trajectories, because the required KF-structuring isn't there" (Maddy, 2007, p. 272). From the perspective I advocate here, what Maddy is highlighting is that a logic must "fit" its domain of application, otherwise it will not effectively serve as a tool for valid inferences.

Setting aside Maddy's and Sher's realist perspectives, what we learn from their accounts is that a logic has to represent the formal properties (in Sher's sense) of its domain of application to function effectively. Both are concerned with logics intended for broad applications, which is why they focus on very general structural properties. However, the request for a wide scope of application can be seen as a deliberate design choice. If one aims to design a logical tool for use in very specific and limited domains, properties that do not qualify as formal in Sher's sense can be relevant. Recall that Sher defines formal properties as those invariant under one-to-one replacements of *any* individuals. If we restrict this definition to a limited domain *D*, properties invariant under such replacements *within D* suffice to establish logical principles and rules of inference *within D*. This means that, in principle, if the intended domain of a logic is confined to natural numbers, the property of being even or odd could ground its mechanisms of inference. Similarly, if the intended domain is confined to people, the property of being an English/British citizen may be logically relevant. Within their respective domains, such regularities ensure truth preservation and are adequately general for their limited purposes.

These are very narrow domains, and it is unlikely that anyone would be interested in setting up logics with so limited applications. But these are just extreme examples of a more pervasive phenomenon. Typically, we lack general-purpose tools capable of universal application. Instead, our toolkit is filled with a variety of instruments designed to assist with specific tasks in specific situations. Accordingly, we have logical tools proper to facilitate the production and recognition of valid inferences about the set-theoretic universe, for example, but not fully equipped to handle other subjects, such as the subatomic universe (Krause, 2023). In the perspective I advocate here, the relationship between logics and their domains of application is analogous to that between screwdrivers and screws: different types of screw heads (Philips, slotted, hexagonal, etc.) require different types of screwdrivers. There is no universal logic, just as there is no universal screwdriver. The logic (screwdriver) one employs depends on the intended domain of application (the type of screw one aims to drive).

Different domains of application may have different inhabitants and, therefore, possess different structural properties. Some domains may exclusively include constructive objects, while others may also include non-constructive objects. Some domains may feature objects with vague properties, while others may consist solely of clear-cut objects, and still others may contain non-individuals (Krause, 2023). Furthermore, there is no barrier preventing a logician from designing a logical tool intended for a domain that does not really exist.[4] What is common to all logics is the necessity to represent certain aspects of their domains of application, irrespective of the nature of these domains.

This view clearly supports pluralism, particularly if we acknowledge that the domain of application of a logic does not necessarily have to correspond to existing reality. Merely hypothetical domains, governed by idealized or entirely fictional structural laws, can give rise to a variety of logics. This partially explains the relative freedom logicians have when creating logics. A mere alteration in the structure of the domain (or in how we conceptualize this structure) can lead to a new logic. However, unlike Carnap's Principle of Tolerance, the point here is not that everyone is free to build up his own logic as they wish. Instead, once a domain of application is selected and its structural properties are well-defined, various representational constraints come into play. Freedom resides mostly (but not solely) in devising and choosing the domain of application.

8 Conventional Aspects of Logics

Now we are in a position to explain why tonk is not an acceptable logical operator. As discussed above, logical laws must capture the structural aspects of their domains of application that ensure transmission of truth. These structural aspects are what ensure that, if a given state of affairs (the one represented by the premises) is the case, then another state of affairs (the one represented by the conclusion) is also the case. The rules governing tonk allow for the inference of any arbitrary Q from any arbitrary P. This means that they are not capturing any structural aspect of the most usual domains of application of logic. In essence, by permitting the inference of any Q from any P, they are blind to any relationships that may exist between states of affairs. In plain words, they are *false* about the structural relationships that typically

[4] The ability of logicians to create logics regardless of the existence of their intended domain of application should not be viewed as a contentious assertion. For instance, while philosophers have not yet reached a consensus on the existence of the set-theoretic universe, this does not hinder logicians from designing logical tools for this domain.

exist in usual domains of application of logic. They could only be true of a rather unusual domain in which any state of affairs follows any other.

Tonk nicely exemplifies the type of conventional freedom that is restricted by the requirement that logical tools must align with (represent) their domains of application. However, as we will see in this section, even when a perfect alignment is achieved, there still remains room for conventions in logic.

As defined in Sect. 4, following (Azzouni, 2014), a practice is conventional when it is one course of action selected from various options that are roughly equal in terms of efficiency and effectiveness. Efficiency and effectiveness are measured against an objective. Thus, whether a tool's aspect is conventional or mandatory must be adjudicated according to the intended objective. Conventional aspects are those for which, given an objective, there are multiple ways to achieve it, each of them being equally successful. In other words, when multiple solutions for a problem are available, there is room for convention. In such cases, the choice of one solution over the others is up to personal preferences, secondary objectives, tradition and possibly other factors. Mandatory aspects, in turn, are those for which there is no alternative.

When it comes to deductive logic, the identification of conventional and mandatory aspects is made, in general, considering its primary objective, namely, truth preservation in the intended domain of application. If there are multiple solutions for the problem of ensuring truth preservation in the intended domain, there is room for choice and, therefore, for the adoption of conventions. Naturally, other objectives may come into play and what was conventional with regard to truth preservation only may become mandatory with regard to truth preservation *plus* additional objectives. Conventionality is not absolute, but relative to one's objectives.

The discussion about the representational aspects of logic in the previous section suggests that most of logic's mandatory aspects are representational and, consequently, most of logic's conventional aspects are non-representational. Although this may often be the case, it is not always so. First, because different logical solutions may be anchored in different representational aspects, allowing for the choice of one solution over the others and, consequently, the choice of the representational aspects to be considered. Second, because technical constraints may require certain non-representational aspects to be mandatory. Thus, what is decisive for the existence of conventions in logic is the availability of multiple solutions to a problem, regardless of a feature being representational or not.[5]

[5] At this point, it is worth discerning the distinction I am drawing between conventional and representational aspects of logic from the distinction between *representors* and *artifacts* as discussed by Cook (2002). Cook posits that logic's role is to provide models of linguistic phenomena. Models are not precise descriptions, but rather useful ways to represent a phenomenon. They often incorporate additional elements to make it easier to work with. He calls "representors" those aspects of the model that are intended to correspond to real aspects of the phenomenon and "artifacts" those that are not thus intended (Cook, 2002, pp. 236–237). Artifacts in Cook's sense cannot be identified with the conventions I am discussing here. Conventionality here is not defined by the lack of representational content, but rather by the existence of alternative solutions. Furthermore, the instrumentalist view defended here does not commit logic to any modeling purpose; logic includes a representational aspect because it needs to, and not because its primary aim is modeling.

A quick glance at logical literature reveals that, more often than not, there are multiple ways to solve a problem or achieve an objective in logic. As Woods (2023) puts it, "[w]ork on logic in the last couple of decades has started to strongly suggest that the traditional ways of doing logic aren't forced. That is, that we have options about which logic we can use." Woods views logical conventionalism as a type of anti-realism about logic and, in line with the traditional perspective on this matter, he links logical conventionalism with arbitrariness. However, agreeing with the view I am defending here, he acknowledges that arbitrariness in logic does not mean total freedom or unmotivated choices: "The arbitrariness in logic is the availability of alternatives, not the lack of explanation of why we have the logic that we do" Woods, (2023, p. 34, fn. 6).

Examples of alternatives that perform equally well but are not chosen arbitrarily abound in logic. The least controversial ones include the interdefinabilty of logical connectives and different but equivalent axiomatizations of one and the same logic. Insofar as these alternatives are equivalent regarding the problem of ensuring truth preservation in a certain domain, the adoption of one over the others is conventional with respect to this objective. This does not mean, however, that we cannot explain why certain alternatives are preferred. As Woods (2023, p. 34, fn. 6) suggests, "[w]e can explain why we made the particular choices we did in terms of our psychologies, historical accident ..., and such." Another way to explain these choices is by considering secondary objectives. Choosing a particular axiomatization of classical propositional logic, for example, can be driven by the goal of making certain proofs easier. In this case, what was conventional regarding the preservation of truth alone becomes mandatory with respect to preservation of truth plus facilitation of proofs.

Other sources of conventionality—of alternatives—in logic are exemplified by translations of one logic into another, the existence of sublogics, and the recovery of one logic within another. These possibilities suggest that we have a variety of options when choosing which logic to use, as different logics may be equally effective for a specific application. A trivial example: if L_a is a sublogic of L_b, and a certain application requires at least L_a, one may choose between using L_a or L_b. Again, besides tradition and personal preferences, secondary objectives may also explain the choice. One might prefer L_a based on a principle of parsimony, or L_b based on a principle of uniformity (if L_b is required for most applications). Conventionality is relative to objectives, but one is free to choose the objectives one pursues. Translations between logics provide a couple of more illustrative examples.

The existence of a *conservative translation* of a logic L_1 into a logic L_2 shows that valid arguments of L_1 can be translated into valid arguments of L_2 and vice versa (Feitosa and D'Ottaviano, 2001). This means that, considering the sole purpose of making and recognizing valid inferences, it does not matter whether one uses L_1 or L_2. The choice of one or the other is, therefore, a matter of convention.

Pairs of logics that have a conservative translation between them can illustrate that the representational aspects of logic are not always mandatory. For example, Kooi and Tamminga (2013) show that any truth-functional three-valued propositional logic can be conservatively translated into the modal logic S5. Particularly, Kleene's Strong Three-Valued Logic (K3) can be conservatively translated into S5. Kleene created K3 "aim[ing] at getting *tools* that render the analysis of partially defined predicates possible" (Malinowski, 2007, p. 22; emphasis added).[6] To this end, he introduced a third truth-value representing undefiniteness. However, the fact that there is a conservative translation of K3 into S5 demonstrates that it would have been possible to handle partially defined predicates effectively by utilizing S5 and its semantics of possible worlds. K3 and S5 are two logical solutions apparently anchored in different representational aspects. S5 includes the notions of necessity and possibility, interpreted in terms of possible worlds, whereas K3 does not. When dealing with partially defined predicates, anchoring the logic in possible worlds or three truth values is a matter of choice.

Even if Kleene was unaware of S5 as an alternative to K3, its existence demonstrates that preferring K3 over S5 for the intended purpose is conventional. When developing a technical solution to a problem, an engineer may not be aware of any alternatives, but, usually, if there is one solution, there are many. The same principle applies to logic.

Once more, being conventional does not imply being arbitrary or willy-nilly. A preference for S5 in contexts involving alethic modality and K3 in contexts involving the indeterminacy of some propositions may be driven, for instance, by cognitive reasons. These options could be more user-friendly in terms of mental demand. In this case, the tool's suitability for the user is what limits the available options. Furthermore, not only cognitive easiness but also the *weight of precedent* may play a role in such situations (Azzouni, 2014). The first efficient solution to a problem will be preferred simply because it was the first one.

A more striking example that shows how logic's representational aspects are not always mandatory is provided by Kooi and Allard's (2013) demonstration that Priest's Logic of Paradox (LP) can be conservatively translated into S5. At first sight, LP and S5 imply very different pictures of their underlying "realities". LP represents a domain where true contradictions exist, while S5 does not allow true contradictions and includes possible worlds. Despite these radical differences, the existence of a conservative translation between them shows that they function equally well in making and recognizing valid inferences. Considering this sole purpose, thus, preferring LP over S5 or vice versa is a matter of convention. In face of this, it could be suggested that these two logics simply anchor their inference mechanisms in different representational aspects, as K3 and S5 above. However, this is hardly the case, as they imply descriptions of their underlying realities that are not compatible. It is as if they were representing incompatible worlds. There are (at least) two ways of understanding what is going on here.

In the first interpretation, either LP or S5 is a suboptimal tool. Suppose that the intended domain is genuinely contradictory, as dialetheism claims. In this case, LP is the most adequate tool, and S5, although still effective, is deficient in some sense. The case may be like using a screwdriver compared to using a pair of pliers: while the screwdriver fits the screw head better, the pliers can still be used (albeit with more effort) to turn a screw by "pinching" its head and twisting it. Analogously, in

[6] Notice that the use of instrumentalist language to describe logicians' motivations is not unusual.

contexts with true contradictions, LP has a better grip, like the screwdriver, while S5 is compared to pliers: it does not fit the domain as well but may be used for the same purpose because it still manage to grip in some way, albeit with some compromise in usability. And vice versa, if the intended domain is genuinely non-contradictory.

Another way of interpreting the disparity of representational content between LP and S5 in the face of the existence of a conservative translation between them is to view this as an instance of a more general phenomenon, also present in mathematics. Putnam describes the mathematical phenomenon as follows:

> In my view the chief characteristic of mathematical propositions is the very wide variety of equivalent formulations that they possess. I don't mean this in the trivial sense of cardinality: of course, every proposition possesses infinitely many equivalent formulations; what I mean is rather that in mathematics the number of ways of expressing what is in some sense the same fact (if the proposition is true) while apparently not talking about the same objects is especially striking. (Putnam, 1967, p. 7)

The existence of a conservative translation between LP and S5 might show that in logic, too, there are multiple ways of expressing the same things (in this case, judgments about which inferences are valid) "while apparently not talking about the same objects." But how can this be possible? Recall that, as explained in Sect. 7, the domain of application of a logic does not necessarily have to correspond to existing reality. If this is the case here, the standard interpretations of LP and S5 can be seen as including useful fictions created as part of the logical solutions, without necessarily reflecting an external reality. If this is so, mentions of possible worlds and true contradictions should be viewed as tools for understanding, rather than as precise representations.

Before concluding, another source of conventionality in logic is worth mentioning: "theory choice." Anti-exceptionalists about logic have suggested that logicians, similar to scientists in general, rely on abductive methods for theory selection. If this is so, logicians have to face the fact that, most times, abduction is unable to select just one theory. Usually, different theories fare equally well the abductivist test, as (Woods, 2023, p. 33) describes:

> If we take seriously …that we ought to decide which logic to accept or whether to revise our own in light of various theoretical virtues, we need to acknowledge that there can be ties in how virtuous various logical theories are. And a quick inspection of the literature reveals that this is the case: various ways of addressing paradoxes, comparisons between various ways of extending our logic to deal with higher mathematics, and many ways of addressing issues with classical logic all seem to stack up more or less on a par with respect to intuitive criteria of theory choice. So it seems we can faultlessly have a different logic.

Thus, as far as abduction—or a similar method for tool choice—is the method of theory choice in logic, typically there are multiple alternatives and there are no definitive criteria to narrow down the range of viable alternatives to just one. If this is so, logicians' choices and preferences, in these cases, are conventional.

We can see so much room for conventions in logic only when we understand that conventionality is defined by the existence of alternatives, rather than by total creative freedom and subsequent mutual agreement on an arbitrary basis. In the understanding of conventionality defended here, mutual agreement still plays a role when logicians jointly decide to favor an alternative over others, but this is not what introduces conventionality in the first place. Mutual agreement would not be necessary if there were not, previously, multiple alternatives available.

9 Logical Instrumentalism

At this point, it is clear that the position defended here belongs to the camp of logical instrumentalism. In this section, I clarify my version of logical instrumentalism by comparing it with Haack's (1978) and Commandeur's (2023) descriptions thereof.

Haack's account of logical instrumentalism captures what is perhaps the most common view on this matter. According to her, instrumentalism is the view that "there is no 'correct' logic" because "the notion of correctness is inappropriate" when it comes to evaluating logics (Haack, 1978, p. 221), "though it might be conceded that it is appropriate to speak of one system's being more fruitful, useful, convenient ...etc. than another (perhaps: for certain purposes)" (Haack, 1978, p. 224). This is so because the instrumentalist believes, according to Haack, that there is no external parameter by which a logical system can be judged as correct or incorrect. "An instrumentalist will only allow the 'internal' question, whether a logical system is *sound*, whether, that is, all and only the theorems/syntactically valid arguments *of the system* are logically true/valid *in the system*" (Haack, 1978, p. 224). Commandeur (2023, p. 4) aptly summarizes Haack's definition of instrumentalism as involving two theses:

> **Logic-as-Tool View**: logics are best understood as tools, a technology or instruments to achieve particular goals or purposes.
> **Non-Representationalism about Logic**: logics do not represent any extra-systemic phenomenon.

The instrumentalist view I present in this chapter does not align with these two theses. Although I endorse the logic-as-a-tool view, I reject non-representationalism about logic. This rejection is related to a problem Commandeur identifies in the traditional conception of logical instrumentalism as described by Haack. The problem is that traditional instrumentalism is unable to explain why a logical system is superior to others for a specific application. According to Commandeur's diagnosis, this difficulty arises from instrumentalism's non-representationalist stance.

> Recall that non-representationalism about logic is the view that logics do not represent any extra-systemic phenomenon. However, especially when combined with the logic-as-tool view that emphasizes the success of logical systems on a variety of applications, non-representationalism faces the challenge of accounting for precisely the success of particular logical systems. And without an explanation of the success of the logical system, logical systems might in some cases indeed turn out to be useful tools for particular applications, but since their success is ...apparently only due to magic or happenstance, we are offered no significant or 'philosophically illuminating' insights. For example, it cannot account for the fact that arguments of a certain form consistently turn out to preserve truth. (Commandeur, 2023, p. 10)

The view I am defending here agrees with Commandeur's diagnosis and overcomes traditional instrumentalism's deficiency by evoking the representational aspects of logical tools. As we have seen, tools partly owe their success to the representational aspects they incorporate. The superiority of a logical tool for a particular application is partly explained by how well it fits the domain of application. Fitness is achieved by incorporating representations of elements from the domain into the logic.

Having rejected traditional logical instrumentalism as described by Haack, Commandeur proposes an alternative view on the subject. His version of instrumentalism also departs from Haack's by rejecting non-representationalism about logic. However, Commandeur goes further and argues that logical instrumentalism should be inspired by scientific instrumentalism. In Commendeur's opinion "by no means does the scientific instrumentalist deny there to be particular phenomena 'out there' that our scientific theories aim to 'capture'" (Commandeur, 2023, pp. 14–15). Accordingly, he suggests that the logical instrumentalist should acknowledge that logic aims to explain extra-systemic phenomena as well. These phenomena, though, do not concern validity itself, "but rather the phenomena for which we need the notion of validity to account for" (Commandeur, 2023, p. 18). "On this picture, what logicians do is setting up a theory, that introduces the technical notion of validity, that explains why particular inferences are truth-preserving. As such, they come along with a *model* for particular phenomena, namely particular truth-preserving inferences" (Commandeur, 2023, p. 18). In this way, Commandeur links his instrumentalism with the logic-as-model view, as proposed by Cook (2002) and Shapiro (2014).

My version of logical instrumentalism contrasts with Commandeur's in that I do not believe that logic aims to model or explain any phenomenon. In the view defended here, the primary purpose of logic is to create tools that help in producing and recognizing inferences that preserve truth. To achieve this, logical tools need to include representations, but these representations do not depict "phenomena for which we need the notion of validity to account for." Rather, they mirror aspects of the intended domain of application, such as the fact that the domain is organized in objects, which have properties, and so on, as explained in Sect. 7. Logicians do not postulate the concept of validity to explain these facts. Rather, in my view, they create this concept as part of a technical solution to the problem of generating and recognizing truth-preserving inferences. In this account, validity is a concept defined within the system (agreeing with Haack's account), even if its definition needs to consider external factors due to technical requirements. As I have proposed elsewhere (dos Santos, 2021b), logic is more similar to technology or engineering than to science. The design of an airplane must take into account facts about the atmosphere, but it is by no means a *model* of atmospheric phenomena. Furthermore, as discussed in Sect. 8, the incorporation of specific representational aspects in logic is up to design choices. The representational elements incorporated may differ based on the requirements of the envisioned solution.

10 Concluding Remarks

This chapter begins by raising the question of whether logics are true and, if so, true of what. My response to this question is that logics are cognitive tools. Typically, tools are not representational artifacts and, as such, do not qualify as true or false. Tools are crafted to serve practical purposes and, as such, are better characterized as useful or useless, efficient or inefficient, and the like. However, we have seen that tools must incorporate representational aspects to ensure their functionality. In this broader sense, we can still speak of truth or falsity in tools.

It may seem unusual and peculiar to describe the tip of a screwdriver as true or false in relation to the head of a screw. Nonetheless, the idea is that there must be a correspondence between them for the screwdriver to fulfill its purpose. As for logical tools, the idea is the same, although the peculiarity of ascribing truth to a tool is mitigated by the shared misperception of logical tools as theories (which are natural truth bearers). Logical truth, in this context, lies in the correspondence between the inference mechanisms of a logical tool and the relevant aspects of its domain of application.

A cautionary remark is in order. As mentioned above, what I am describing as the domain of application of a logic does not necessarily have to correspond to an existing reality. We have the freedom to create a domain along with its structural laws—whether it's an idealization of a limited sector of reality, a wholly abstract structure, or even a fictional construct. Furthermore, even when the intended domain is conceived as an independent reality, there can be divergent views regarding the laws governing it (e.g., intuitionists and classicists differ on the laws governing the mathematical universe). Accordingly, logical truth understood in these terms does not equate to factual truth. It simply signifies the existence of a "match" between a mechanism of inference and the perceived properties of a topic or subject to which that mechanism is intended to be applied. Whether these perceived properties are authentic, mere idealizations, or fictions is irrelevant from the perspective of the logical tool.

These observations provide a way of understanding logic's neutrality with respect to the ontology and the metaphysics of the universe. From the perspective of a logic, it is immaterial how the intended domain of application *really* is. A logical tool is never applied to reality itself, but solely to our linguistic representations of it. The design of a logical tool starts from a linguistic specification of its domain, and logical truth is confined to the correspondence between the logic and that specification. The verification of the accuracy of the specification is not a part of the process of constructing a logical tool.

This explains the freedom logicians have to develop logics that are not primarily intended for real-life applications. Much of the work in logic is dedicated to enhancing the range of logical tools, often without a specific application in mind. Logicians' interests may lie exclusively in the creation of conceptual tools. From such a diverse conceptual toolbox, natural scientists, computer scientists, linguists, and philosophers have plenty of options to choose the tools that best suit their needs.

Furthermore, logicians may not only focus on creating new logical tools but also on examining the characteristics of inferential systems in general. This aspect of logicians' work is akin to an engineer investigating the properties of machines in general, such as material resistance and friction. Additionally, both an engineer and a logician may create a tool purely for the purpose of exploring technical alternatives.

All in all, the perspective defended in this chapter is that logic is less like a scientific field, in which researchers are concerned with describing specific phenomena, and more like a technological field, where researches are concerned with studying technical solutions and developing tools.

References

Azzouni, J. (2014). A defense of logical conventionalism. In P. Rush (Ed.), *The metaphysics of logic* (pp. 32–48). Cambridge University Press.

Benacerraf, P. (1983). Mathematical truth. In P. Benacerraf & H. Putnam (Eds.), *Philosophy of mathematics: Selected readings* (pp. 403–420). Cambridge University Press.

Carnap, R. (1983). Empiricism, semantics and ontology. In P. Benacerraf & H. Putnam (Eds.), *Philosophy of mathematics: Selected readings* (pp. 241–257). Cambridge University Press.

Carnap, R. (2001). *Logical syntax of language*. Routledge.

Commandeur, L. (2023). Logical instrumentalism and anti-exceptionalism about logic. *Erkenntnis*, 1–21.

Cook, R. T. (2002). Vagueness and mathematical precision. *Mind, 111*(442), 225–248.

Dascal, M. (2002). Language as a cognitive technology. *International Journal of Cognition and Technology*, 35–61.

dos Santos, C. F. (2021a). Enculturation and the historical origins of number words and concepts. *Synthese, 199*, 9257–9287.

dos Santos, C. F. (2021b). Intuitions, theory choice and the ameliorative character of logical theories. *Synthese, 199*, 12199–12223.

dos Santos, C. F. (2025). *Numbers as cognitive tools: An empirically informed nominalistic account of the nature of numbers*. Springer.

dos Santos, C. F. (2023). Non-numerical methods of assessing numerosity and the existence of the number sense. *Journal of Numerical Cognition, 9*(2), 363–379.

Dutilh Novaes, C. (2012). *Formal languages in logic: A philosophical and cognitive analysis*. Cambridge University Press.

Dutilh Novaes, C. (2021). *The dialogical roots of deduction: Historical, cognitive, and philosophical perspectives on reasoning*. Cambridge University Press.

Evans, J., Newstead, S., & Byrne, R. (1993). *Human reasoning: The psychology of deduction*. Psychology Press.

Feitosa, H. A., & D'Ottaviano, I. M. L. (2001). Conservative translations. *Annals of Pure and Applied Logic, 108*(1–3), 205–227.

Frege, G. (1960). *The foundations of arithmetic: A logico-mathematical enquiry into the concept of number*. Harper & Brothers.

Frege, G. (1997a). Begriffsschrift. In M. Beaney (Ed.), *The Frege reader* (pp. 47–78). Blackwell.

Frege, G. (1997b). Thought. In M. Beaney (Ed.), *The Frege reader* (pp. 325–345). Blackwell.

Haack, S. (1978). *Philosophy of logics*. Cambridge University Press.

Hjortland, O. T. (2017). Anti-exceptionalism about logic. *Philosophical Studies, 174*(3), 631–658.

Hlobil, U. (2021). Limits of abductivism about logic. *Philosophy and Phenomenological Research, 103*, 320–340.

Kant, I. (1998). *Groundwork of the metaphysics of morals*. Cambridge University Press.

Knops, A. (2020). *Numerical cognition: The basics*. Routledge.
Kooi, B., & Tamminga, A. (2013). Three-valued logics in modal logic. *Studia Logica, 101*(5), 1061–1072.
Krause, D. (2023). On identity, indiscernibility and (non-)individuality in the quantum domain. *Philosophical Transactions of the Royal Society A, 381*(2255).
Maddirala, N. (2014). *Philosophy of logical practice: A case study in formal semantics* [Master's thesis. Universiteit van Amsterdam].
Maddy, P. (2007). *Second philosophy: A naturalistic method*. Oxford University Press.
Maddy, P. (2014). *The logical must: Wittgenstein on logic*. Oxford University Press.
Malinowski, G. (2007). Many-valued logic and its philosophy. In D. M. Gabbay & J. Woods (Eds.), *Handbook of the history of logic* (Vol. 8). Elsevier.
Martin, B., & Hjortland, O. T. (2020). Logical predictivism. *Journal of Philosophical Logic*, 1–34.
Martin, B., & Hjortland, O. T. (2023). Evidence in logic. In M. Lasonen-Aarnio & C. M. Littlejohn (Eds.), *Routledge Handbook of the Philosophy of Evidence*. Routledge.
Menary, R. (2007). *Cognitive integration: Mind and cognition unbounded*. Palgrave Macmillan.
Menary, R., & Gillett, A. (2022). The tools of encultaration. *Topics in Cognitive Science, 14*, 363–387.
Overmann, K. A. (2023). *The materiality of numbers: Emergence and elaboration from prehistory to present*. Cambridge University Press.
Pohl, R. F. (Ed.). (2017). *Cognitive illusions: Intriguing phenomena in thinking, judgment and memory* (2nd ed.). Routledge.
Priest, G. (2006). *Doubt truth to be a liar*. Clarendon Press.
Priest, G. (2023). Logic as applied mathematics: with particular application to the notion of logical form. *Logic and Logical Philosophy, 32*, 443–457.
Prior, A. N. (1960). The runabout inference-ticket. *Analysis, 21*(2), 38–39.
Putnam, H. (1967). Mathematics without foundations. *The Journal of Philosophy, 64*(1), 5–22.
Sá, W. C., West, R. F., & Stanovich, K. E. (1999). The domain specificity and generality of belief bias: Searching for a generalizable critical thinking skill. *Journal of Educational Psychology, 91*(3), 497–510.
Schmitt, R. (1965). Heidegger's analysis of 'tool'. *The Monist, 49*(1), 70–86.
Shapiro, S. (2006). *Vagueness in context*. Clarendon Press.
Shapiro, S. (2014). *Varieties of logic*. Oxford University Press.
Sher, G. (2016). *Epistemic friction: An essay on knowledge, truth, and logic*. Oxford University Press.
Sher, G. (2011). Is logic in the mind or in the world? *Synthese, 181*, 353–365.
Stanovich, K. E. (2003). The fundamental computational biases of human cognition: Heuristics that (sometimes) impair decision making and problem solving. In J. E. Davidson & R. J. Sternberg (Eds.), *The psychology of problem solving, page* (pp. 291–342). Cambridge University Press.
Stokhof, M., & van Lambalgen, M. (2011). Abstractions and idealisations: The construction of modern linguistics. *Theoretical Linguistics, 37*(1/2), 1–26.
Warren, J. (2020). *Shadows of syntax: revitalizing logical and mathematical conventionalism*. Oxford University Press.
Warren, J. (2015). Talking with tonkers. *Philosophers' Imprint, 15*(24), 1–24.
Woods, J. (2023). A sketchy logical conventionalism. *Aristotelian Society Supplementary Volume, 97*(1), 29–46.

Logical Abductivism and the Liar Paradox

Elke Brendel

Abstract Logical abductivism is a specific version of an anti-exceptionalist view in the philosophy of logic. According to logical abductivism, logical theories are justified, assessed or revised via scientific methods of rational theory choice. The paper proposes four criteria that typically guide an abductive cost-benefit analysis when comparing rival logics. These criteria are *non-triviality*, *adequacy with evidence or data*, *appropriate balance between strength and simplicity*, and *intertheoretic coherence*. They will be used as a guideline to examine, in particular, Timothy Williamson's argument that even in light of the semantic paradoxes, classical logic is abductively preferable to its non-classical alternatives, such as the logics K_3 and LP. Against Williamson's view, it is contended that scientific reasoning does not universally and categorically employ the rules of classical logic. Moreover, many classical solutions to the Liar paradox are burdened with abductive costs. It is also argued that the concern, according to which non-classical gap- or glut-theories of truth would have an unfavorable effect on epistemology, is unjustified. In the last part of the paper, an *operator account of truth* is suggested and abductively compared to predicate theories of truth.

1 Anti-exceptionalism About Logic and Logical Abductivism

Anti-exceptionalism about logic has received much attention in contemporary philosophy of logic. Generally, and broadly speaking, logical anti-exceptionalism conceives of logic as, in a certain sense, *continuous with science*. In an often-quoted statement by Ole Hjortland, logical anti-exceptionalism is expressed as follows:

> Logic isn't special. Its theories are continuous with science; its method continuous with scientific method. Logic isn't a priori, nor are its truths analytic truths. Logical theories are revisable, and if they are revised, they are revised on the same grounds as scientific theories. These are the tenets of anti-exceptionalism about logic. (Hjortland, 2017: 631)

E. Brendel (✉)
Institut für Philosophie, Universität Bonn, Bonn, Germany
e-mail: ebrendel@uni-bonn.de

S. Molick (ed.), *Demarcating Logic and Science*, Studies in Applied Philosophy, Epistemology and Rational Ethics 80, https://doi.org/10.1007/978-3-032-25674-4_2

Logical anti-exceptionalism comprises several metaphysical, epistemic, methodological and normative views about logic which are, in different ways and to different degrees, in opposition to a *traditional account of logic* (see Martin and Hjortland, 2022).

Logic is traditionally conceived as an *a priori* discipline whose principles are independently justified from empirical evidence. Many proponents of a traditional account of logic contend that basic logical laws are known either via some kind of rational insight (see, for example, Bealer, 1998) or by mere reflection on the meaning of central logical terms (see, for example, Ayer, 1952: ch. IV). Furthermore, the logical inference rules are traditionally understood as purely *formal* and *necessary* principles that reflect the most *fundamental* and *general* structures of human reasoning (see, for example, Frege, 1893: xv; Kant, 1781: A52/B76; Tarski, 1936: 414). In contrast to this traditional account of logic, anti-exceptionalists (see, for example, Bueno, 2010; Bueno and Colyvan, 2004; da Costa and Arenhart, 2018; Hjortland, 2017; Maddy, 2002; Priest, 2014, 2016; Russell, 2014; Williamson, 2017) believe that logic does not enjoy a special and exceptional status among the sciences. Those who endorse *epistemological* forms of logical anti-exceptionalism argue that the ways we justify, select and revise our logical principles and theories are comparable to the methods used in the empirical sciences. Some of them even envisage themselves in a situation similar to that described by Otto Neurath's boat and advocate a kind of *holistic justification* of logical principles:

> The logical boat, as much as the empirical one, must be reconstructed at sea, plank by plank, with no privileged and with no pre-ordained starting-point. (Sainsbury, 2002: 14)

With regard to the epistemology of logic, logical anti-exceptionalists typically subscribe to an *abductive methodology*. They contend that not only in the sciences but also in logic the rational processes of the justification, evaluation and selection of theories are based on *abductive reasoning*. For logical abductivists, logical theories, just as scientific theories, are rationally assessed with respect to how well they account for relevant evidence and data and how well they satisfy certain scientific virtues, such as explanatory and unifying power, avoidance of adhocness, fruitfulness, etc. On the basis of an abductive argument, i.e., via some kind of an *inference to the best explanation*,[1] logicians choose the theory that best accommodates those theoretical criteria and scientific virtues.[2]

There are, however, many aspects of an abductive methodology in the epistemology of logic on which logical abductivists disagree. First of all, they disagree on the question of what the most important abductive criteria in logical theory choice are and how those criteria are to be weighted. Second, logical abductivists disagree about the epistemic role *paradoxes* play in the justification of logical theories. In

[1] Although in the case of logic the explanation is typically *not* a *causal* explanation (see Williamson, 2017: 334).

[2] See, for example, Priest (2014, 2016) and Williamson (2013, 2017) for defending abductive methods in logical theory choice.

particular, some logical abductivists hold different opinions about the adequate logical treatment of the *semantic paradoxes*, especially the *Liar paradox*. In this paper, I will focus on these two aspects of disagreement among logical abductivists.

2 Abductive Criteria for Logical Theory Choice

In what follows, I will give a rough outline of four intuitive criteria that play an important role in logical theory choice. I will elaborate on these criteria only to the extent that is needed to apply them to an abductive cost-benefit analysis of certain logical theories,[3] in particular, with regard to their ability to deal appropriately with the Liar paradox.

The first and most important criterion is *non-triviality*:

Abductive Criterion 1: Non-Triviality
A logical theory T has to be non-trivial. It should not be the case that any sentence can be derived from a logical theory T, i.e., it should be the case that there are sentences σ such that $\nvdash_T \sigma$.[4]

Triviality is the most serious deficiency of a logical theory that cannot be outweighed by other abductive criteria. In proving each sentence and also its negation, a trivial theory can no longer discriminate truths from untruths. Thus, triviality has to be avoided at all costs. Triviality always renders a theory logically and scientifically useless and insignificant.

In the literature on logical abductivism, the criterion of *adequacy with evidence or data* is the most mentioned criterion in logical theory choice (see, for example, Hjortland, 2019: 257; Priest, 2014: 217, 2016: 32; Williamson, 2017: 334f.). This criterion is our second abductive criterion. It can be briefly described as follows:

Abductive Criterion 2: Adequacy with Evidence or Data
A logical theory T should be adequate to the evidence and data that theorists endorsing T aim to account for.

In contrast to the criterion of non-triviality, this criterion is not an absolute and objective one. It depends on the evidence and data that one intends to explain with the logical theory. Furthermore, evidence and data might turn out to be *fallible*. It is also possible that the *criterion of adequacy with evidence or data* can be trumped by other abductive criteria. A logical theory that fits well with evidence and data can nevertheless be rejected in favor of an alternative theory that performs better in the overall abductive balance.

---

[3] In this paper, the term "logical theory" is used in a rather broad sense. A logical theory is a theory about the validity of arguments. So, proponents of different logical theories differ in what they count as a valid argument. The property of validity in a logical theory can be syntactically or semantically formalized in proof-theoretic or model-theoretic ways (see, for example, Hjortland, 2019, pp. 252–256) for an explication of the notion of a logical theory.

[4] "$\vdash_T$" denotes T's derivability relation. Thus, "$\nvdash_T \sigma$" symbolizes that the sentence σ is not provable in T.

What kind of evidence and data should be addressed by a logical theory? Many logical abductivists, like Graham Priest, claim that *intuitions* about *validity* or about the *correctness* of certain *inferences* count as important evidence in evaluating logical theories:

> It is clear enough what provides the data in the case of an empirical science: observation and experiment. What plays this role in logic? The answer, I take it, is our intuitions about the validity or otherwise of vernacular inferences. (Priest, 2016: 41)

However, data based on intuitions of validity might be mistaken or unstable. As Priest emphasizes, intuitions about the correctness of individual logical inferences, especially when applied to concrete everyday situations, are usually clearer and more stable than those about the validity of *general* logical laws:

> If theorization is to take account of such data, they are certainly much softer than those concerning individual inferences. Very often, a form of inference strikes us as correct only because of an impoverished diet of examples. (Priest, 2016: 43)

For example, *modus ponens* appears to be extremely plausible in many ordinary applications. But in light of inferences with possible paradoxical sentences, the intuitive validity of *modus ponens* as a *general* logical inference pattern seems more tenuous.

Besides intuitions about the validity of inferences, intuitions about the meaning, formal structures, and logical behavior of central philosophical concepts (such as *truth*, *modality*, *necessity*, *possibility*, *probability*, *knowledge*, *belief*, *identity*, *essence*, *quantification*, etc.) or logical constants (as, for example, *negation* or *implication*) play an important evidential role in assessing logical theories. For example, in the case of modal logic, we find, as Timothy Williamson claims, "no case in which a material conditional is necessary, and so is its antecedent, but its consequent is not" (Williamson, 2013: 425). This data gives rise to the corresponding axiom schema $K : \Box(p \rightarrow q) \rightarrow (\Box p \rightarrow \Box q)$. So, K explains a fundamental pattern in modal data by bringing it under a universal law, and, thus, unifies and generalize this modal evidence. But even proposed laws of modal logic based on patterns of modal data, such as K, must be weighed against other abductive criteria. They have to be "integrated into a total modal logic, such as S5, with the required theoretical virtues" (Williamson, 2013: 426).

Sometimes also *empirical* or *observational data* play an evidential role in logical theory choice. For example, Hilary Putnam, famously argued for *quantum logic* in which, in particular, the classical law of distributivity fails in order to formally represent certain quantum physical effects (see Putnam, 1969).

The *third* important abductive criterion of logical theory assessment is the following criterion of an *appropriate balance between strength and simplicity* of a logical theory:

> **Abductive Criterion 3: Appropriate Balance Between Strength and Simplicity**
> A logical theory T should be logically strong enough (and not unnecessarily too strong) in order to be as informative and explanatory as possible with regard to T's intended goals and area of application.

T should be as simple as possible in order to formulate logical rules that are sufficiently general and universal with regard to T's intended goals and area of application. Unnecessary restrictions or exceptions to the generality of the logical rules should be avoided.

The expressive and explanatory strength of a logical theory T required by this criterion is not meant to be identical with T's deductive or logical strength. Deductive strength of a logical theory is usually defined in terms of its consequence relation. For example, a logical theory T_1 is deductively stronger than a theory T_2 if all logical consequences of T_2 are also logical consequences of T_1, but not *vice versa*. Gillian Russell has pointed out that logical strength *per se* is "neither a virtue nor a vice" (Russell, 2019: 555). Clearly, a trivial theory is too strong since it allows to prove everything. Furthermore, as we will see in Sect. 4, the paraconsistent logic LP and a K_3-based gap-theory, are deductively weaker than classical logical theories. But this doesn't necessarily mean that a classical theory is therefore abductively preferable to LP and K_3. Sometimes, as Hjortland noticed, a deductively weaker logical theory can be favourable to a stronger theory because a weaker theory might allow for more fine-grained discriminations (see Hjortland, 2017: 646). Thus, it seems possible that a logical theory T_1 which is deductively weaker than a theory T_2, could still be explanatory stronger than T_2. Its greater explanatory strength could then provide an abductive ground to prefer T_1 over T_2.

The importance of adequate expressive power as an abductive criterion of logical theory choice can be seen, in particular, in the negative attitude towards so-called *revenge problems* in the discussion about appropriate solutions to semantic paradoxes. As it will be argued in Sect. 5 in more detail, revenge problems occur when the measure taken to resolve the original Liar paradox can itself be used to create new paradoxes. Those "revenge paradoxes" are typically formulated via certain central meta-theoretic notions, such as notions from the model theory of the given logic. Many revenge paradoxes can therefore only be circumvented by excluding those revenge-paradox-generating notions from the object-language of the logical theory in question—which often results in a significant limitation of the theory's expressive power. So, a theory's capability of treating its revenge problems in a way that does not result in severe expressive limitations, can contribute to the abductive advantages of this theory.

The *fourth* abductive criterion of logical theory choice concerns the *coherence* of a logical theory T with other theories that are relevant to T in certain respects:

Abductive Criterion 4: Intertheoretic Coherence

A logical theory T should cohere with central claims and tenets of other theories to which T is related or with other potential areas of applications of T.

The criterion of *intertheoretic* coherence stresses the point that an overall abductive evaluation of a logical theory should not only consider how well this theory fares with regard to the expressive and explanatory adequacy of the theory's most central inner-theoretic aims but should also assess the theory's adequacy in a more global way. A logical theory whose central principles and theorems are in contrast with well-established results of other philosophical or scientific theories suffers from intertheoretic incoherence—in particular, if these philosophical or scientific theories are among the potential scope of application of the logical theory. For example, a logical theory T lacks intertheoretic coherence if T's formal account of truth doesn't provide an adequate basis for dealing with other concepts which essentially entail the concept of truth, such as *necessity*, *knowledge* or *probability*.

A theory's intertheoretic coherence is closely connected to the theory's explanatory adequacy. However, since intertheoretic coherence plays a particularly important role in logical abductivism, as we will see, for example, when examining Williamson's or Halbach's abductive argument in favour of classical logic, this virtue is worth mentioning as a separate abductive criterion of logical theory choice.

These four criteria of *non-triviality*, *adequacy with evidence or data*, *appropriate balance between strength and simplicity*, and *intertheoretic coherence* are neither meant to be exhaustive nor exclusive. The criterion that has the highest abductive weight is *non-triviality*. A logical theory that allows to prove everything can never be chosen as the best logical theory. The other criteria leave more room for different assessments in abductive logical theory choice. How the criteria of *adequacy with evidence and data*, *appropriate balance between strength and simplicity*, and *intertheoretic coherence* are understood and which abductive weight they carry in specific cases of logical theory choice can be the subject of controversial debate.

In the following, the four outlined abductive criteria of logical theory choice will be applied to the debate about the most appropriate solution to the *Liar paradox*.

3 The Liar Paradox in Classical Logic

Although Williamson admits that "[a]t first sight, the semantic paradoxes constitute unusually promising ground for an abductive critique of classical logic" (Williamson, 2017: 339), he nevertheless argues that in assessing logic abductively, classical logic still fares better than any of its non-classical rivals. Before we examine Williamson's abductive argument for classical logic, let us first recapitulate under what conditions classical logic can lead to semantic paradoxes, such as the Liar paradox.

In the following, we consider a theory in classical logic whose first-order language allows encoding its own syntax via *Gödelization*, such as the system of Robinson arithmetic Q. Furthermore, we add to the system Q both a *truth predicate* and an *unrestricted disquotation schema*. This enriched system of Robinson arithmetic will be called Q^+.

Just as Q, Q^+ provides *names* for all expressions of its language. In particular, if ζ is a formula of the language of Q^+, $\langle\zeta\rangle$ is a *term* of the language of Q^+, where "$\langle\zeta\rangle$" stands for the numeral of the Gödel number for ζ—the *quotation name* of ζ. Among the predicates of Q^+'s language, there is a special predicate "Tr" which functions as a *truth predicate* within Q^+. If ζ is a formula of Q^+'s language, "$Tr\langle\zeta\rangle$" denotes the truth of ζ. Furthermore, and most importantly, Q^+ additionally contains the following *T-schema* (TS), which governs the disquotational feature of the truth predicate Tr:

$$\text{TS}: \vdash_{Q^+} Tr\langle\sigma\rangle \leftrightarrow \sigma \quad \text{(for all sentences } \sigma \text{ of the language of } Q^+\text{)}.$$

TS expresses a minimal adequacy condition of a truth predicate since it guarantees the material equivalence of a sentence σ with a sentence stating the truth of σ. TS can be separated into the following inference rules:

capture: $\sigma \vdash_{Q^+} Tr\langle\sigma\rangle$
release: $Tr\langle\sigma\rangle \vdash_{Q^+} \sigma$
(for all sentences σ of the language of Q^+).

It is a well-known fact that Q^+ is *provably inconsistent*. Because in Q^+ the principle *ex contradictione quodlibet* (ECQ) holds, the inconsistency of Q^+ leads to the *triviality* of Q^+.

The proof of Q^+'s triviality makes use of the so-called *diagonal lemma*, which holds for a logical system of Robinson arithmetic such as Q^+. According to the diagonal lemma, there is, for any formula $B(y)$ containing just the variable y free, a sentence σ such that: $\vdash_{Q^+} \sigma \leftrightarrow B(\langle\sigma\rangle)$, where "$B(\langle\sigma\rangle)$" results from "$B(y)$" by replacing all free occurrences of "y" in "$B(y)$" with "$\langle\sigma\rangle$". As a consequence, the diagonal lemma allows constructing a sentence λ such that: $\vdash_{Q^+} \lambda \leftrightarrow \neg Tr\langle\lambda\rangle$, where $\neg Tr\langle\lambda\rangle$ results from the formula "$\neg Tr(y)$" (with just the free variable y) by replacing the free occurrence of "y" in "$\neg Tr(y)$" by "$\langle\lambda\rangle$", i.e., by the quotation name of λ. The construction of such a sentence λ uses the *diagonalization*[5] on $\neg Tr(y)$ which, intuitively, is a sentence "saying" of itself that it is not true. Since in a classical bivalent theory the negation of the truth of a sentence σ and the falsity of σ are the same, λ also "says" of itself to be false. Thus, λ formally represents within Q^+ the famous Liar sentence.

Here is a version of the proof of Q^+'s triviality[6]:

1( )	$\lambda \leftrightarrow \neg \mathrm{Tr}\langle\lambda\rangle$	(by the diagonal lemma)
2(2)	$\mathrm{Tr}\langle\lambda\rangle$	(assumption)
3(2)	λ	(2, release)
4(2)	$\neg \mathrm{Tr}\langle\lambda\rangle$	(1, 3, $\leftrightarrow$-elimination)
5(2)	$\mathrm{Tr}\langle\lambda\rangle \wedge \neg \mathrm{Tr}\langle\lambda\rangle$	(2, 4, $\wedge$-introduction)
6( )	$\neg \mathrm{Tr}\langle\lambda\rangle$	(2, 5, *reductio ad absurdum*)
7( )	λ	(1, 6, $\leftrightarrow$-elimination)
8( )	$\mathrm{Tr}\langle\lambda\rangle$	(7, capture)
9( )	$\mathrm{Tr}\langle\lambda\rangle \wedge \neg \mathrm{Tr}\langle\lambda\rangle$	(6, 8, $\wedge$-introduction)
10( )	β	(9, ECQ, where β is an arbitrary sentence)

As argued above, triviality is the most serious deficiency of a logical theory. A logical theory that does not fulfill the abductive criterion of non-triviality must be dismissed. The above proof indicates that, in order to avoid triviality, one cannot keep both all of classical logic and a fully transparent truth predicate that satisfies *capture*

---

[5] For the notion of diagonalization, see, for example, (Boolos et al., 2002: 172).

[6] The numbers in brackets after the line numbers indicate the line numbers of the assumptions from which the sentence in the line depends.

and *release*. Which alternative one should opt in order to satisfy the unconditioned and most highly weighted criterion of non-triviality depends on how the suggested alternatives to Q^+ score with regard to other abductive criteria.

4 Abductive Arguments in Favour of Classical Solutions to the Liar Paradox

According to Williamson, an abductive cost-benefit analysis clearly points in favour of saving classical logic and restricting the disquotational T-schema. Williamson contends that classical logical consequence relations that govern the use of logical constants are integral to the natural sciences and mathematics. The truth predicate and quotation devices are *metalinguistic* and as such of no significant importance for most applications in the sciences and mathematics. Thus, restricting disquotation while saving classical logic is, according to Williamson, abductively favourable. Williamson's abductive argument in favour of classical logic rests, in particular, on a certain application of the abductive criterion of *intertheoretic coherence*. Williamson points out that the abductive methodology "recommends us not to compare two logics only in isolation, but also to compare the results of combining each of them with well-confirmed results from outside logic, such as principles of natural science" (Williamson, 2017: 340). According to Williamson, classical logic coheres much better with most of the natural sciences than any non-classical logic. In particular, as Williamson claims, sciences "that involve some mathematics" use the classical logical constants and a classical consequence relation. So, for Williamson, the most important application of a logic is to establish a logical foundation of the natural sciences. And a logic which fits this role best should obtain priority in a logical theory selection.

In attaching great abductive importance to a logic's capability of cohering with the mathematical theory used by natural scientists, logic, for Williamson, is primarily conceived as a *non-metalinguistic discipline*: logic is about the most general and fundamental structures used by scientists to describe and explain the world. Thus, according to Williamson, the main concern of a logical theory is not a metalinguistic one. Logic does not primarily aim at providing a theory of the semantics and consequence relations of a natural language.[7] That is why Williamson does not attach too much abductive weight to the ability of a logical theory to save an unrestricted disquotation schema, although an unrestricted disquotation schema appears to reflect an important intuitive property of the truth predicate for natural languages. For Williamson, the logical foundation of the natural sciences is the mainly intended area of application of a logic. Since, according to Williamson, the concept of truth and quotation marks are no integral parts of the sciences, restricting the T-schema does not significantly reduce the classical logic's expressive and explanatory power.

[7] For a distinction between metalinguistic and non-metalinguistic theories and the different types of evidence that support those theories, see (Hjortland, 2019: 254f).

Thus, for Williamson, a *classical* logical theory that restricts the T-schema in order to avoid triviality, also satisfies the abductive criterion of *an appropriate balance between strength and simplicity* to a high degree.

The persuasiveness of Williamson's abductive argument in favour of classical logic depends on whether non-classical logics really score worse than classical logic with respect to the abductive criteria. The main accounts of non-classical logic that Williamson is abductively comparing with classical logic, are logics based on a *strong Kleene model*, such as the paracomplete logic K_3, in which each sentence is assigned exactly one truth values from the set $\{1, \frac{1}{2}, 0\}$ and where the value 1 is the only *designated value*, i.e., the value that is preserved by a logical consequence relation.[8] A sentence which is assigned the value 1, can be interpreted as being *true*, whereas a sentence which is assigned the value 0, can be interpreted as being *false*. A sentence that receives the value $\frac{1}{2}$ in a K_3-model, is often interpreted as being "gappy", *undefined* or *neither true nor false*.

Let us now consider a logic K_3 which has, just as Q^+, the resources to encode its own syntax and contains the full T-schema. In such a logic, the Liar sentence λ receives the value $\frac{1}{2}$. It is, therefore, supposed to be a sentence which lacks a classical truth value (see Kripke, 1975). Many classical principles fail in K_3, as, for example, the *law of excluded middle* (LEM).[9] Although ECQ is valid in K_3, the Liar sentence λ does not give rise to triviality since λ does not generate a provable contradiction. In particular, deriving $\neg Tr\langle\lambda\rangle$ via reductio in line 6 from lines 2 and 5 in the classical proof above, would be *incorrect* in a K_3 proof-system. So, the Liar sentence in K_3 does not lead to a provable contradiction, even though the unrestricted T-schema holds.[10]

Another non-classical logic, which, according to Williamson, scores worse than classical logic in an abductive comparison, is Priest's *paraconsistent "Logic of Paradox"* LP (see Priest, 1979, 2008: 122f.). LP is a logical system which contains the full T-schema[11] and allows to encode its own syntax and is also based on a strong Kleene model. The only *formal* difference to K_3 is that in LP the values 1 and $\frac{1}{2}$ are both designated values. A sentence having the value $\frac{1}{2}$ in LP, such as the Liar sentence λ, is a "glutty" sentence, i.e., a sentence being *both true and false*. One

[8] In a strong Kleene model, sentences φ and ψ are assigned the values 0, 1, and $\frac{1}{2}$ in accordance with the following evaluation function V : $V(\neg\varphi) = 1 - V(\varphi)$; $V(\varphi \wedge \psi) = \min(V(\varphi), V(\psi))$; $V(\varphi \vee \psi) = \max(V(\varphi), V(\psi))$; $V(\varphi \rightarrow \psi) = \max(1 - V(\varphi), V(\psi))$.

[9] For "gappy" sentences σ, it holds that $(\sigma \vee \neg\sigma)$ is also a gappy sentence. Thus, a disjunction of the form $(\sigma \vee \neg\sigma)$ does not always have the designated value 1.

[10] It should be noticed that K_3 has no *tautologies*. So, for example, even the *principle of identity* $\sigma \rightarrow \sigma$ is K_3-invalid, since for sentences σ with $V(\sigma) = \frac{1}{2}$, $V(\sigma \rightarrow \sigma) = \frac{1}{2}$, and $\frac{1}{2}$ is not a designated value in K_3. For similar reasons, TS in its *biconditional form* $Tr\langle\sigma\rangle \leftrightarrow \sigma$ is K_3-invalid. However, the *inferential versions* of TS, in form of the rules *capture* and *release*, do hold in K_3. In this sense, one could still claim that an unrestricted T-scheme is maintained in K_3.

[11] In LP, $V(Tr\langle\sigma\rangle \leftrightarrow \sigma) \in \{\frac{1}{2}, 1\}$, i.e., TS is "at least true". Thus, TS is LP-valid (in its biconditional form, but also, of course, in its inferential versions). LP-theorists only reject sentences which are "false only", i.e., sentences σ with $V(\sigma) = 0$. So, an LP-theorist accepts TS. It can, thus, be claimed that the full T-schema is contained and true in LP.

important consequence of LP is that ECQ is *invalid*.[12] In proof-theoretic terms this means that ECQ cannot be used as an inference rule within LP. Thus, in the above proof, the step from line 9 to 10 would be blocked. Although $Tr\langle\lambda\rangle \wedge \neg Tr\langle\lambda\rangle$ is a provable contradiction in LP, this contradictory result does not lead to triviality.

One of Williamson's main arguments against K_3 and LP is that those non-classical theories are inadequate as a suitable foundation for mathematics and the sciences. For Williamson, both K_3 and LP invalidate vast swathes of mathematical and scientific reasoning (Williamson, 2017: 341f.). As we have seen above, LEM fails in K_3. But, as Williamson argues, "mathematicians freely reason in ways that depend on the law of excluded middle" (Williamson, 2017: 341). A K_3-logician, in order to account for the use of LEM in scientific reasoning, has to restrict the instances of LEM to applications in science by "metalinguistic" considerations, "such as that all instances of excluded middle in the language of physics are true" (Williamson, 2017: 341).

Many inference rules of classical logic fail in LP as well. Besides ECQ, there are, for example, also *disjunctive syllogism* or *modus ponens* which are invalid in LP. But these rules are among the most widely used rules in mathematical and scientific reasoning, and in human reasoning in general. Of course, an LP-theorist could permit those classical inference rules for *nonparadoxical* cases. But, according to Williamson, this "piecemeal reintroduction of instances of missing classical principles" (Williamson, 2017: 342) would result in heavy abductive costs:

> Arguably, dialetheists underestimate the costs of abandoning classical logic. Since classical logic is tacitly used throughout science, restricting it has massive knock-on effects. Dialetheists hope to recover classical logic whenever they need it in non-paradoxical situations, on a case-by-case basis, by adding extra assumptions to the effect, "Classical logic is OK in this case". Those assumptions must be added to all sorts of completely normal scientific explanations in non-paradoxical situations to make them work without granting the general validity of classical logic. Messing up elegant scientific explanations with numerous extra assumptions made up as one goes along reduces the power of those explanations all over science. If instead one retains classical logic and makes adjustments elsewhere, one can avoid such widespread losses of explanatory power [...]. (Williamson, 2018: 93)

It is, however, contestable whether scientific reasoning does indeed always follow the rules of classical logic, and if they follow them whether they follow them categorically as universally applicable rules. Do scientists, for example, really argue in line with ECQ or do they always apply a monotonous inference relation? It seems that we often apply classical inference rules only under certain extra-logical assumptions or idealizations (see da Costa and Arenhart, 2018).

Furthermore, so-called *classical recapture* strategies attempt to show that non-classical logicians can use classical logic in certain "safe" contexts, i.e., contexts that did not give them reason to reject classical logic (as, for example, contexts with no paradoxical sentences). For example, Fiore and Rosenblatt have proved a recapture result for K_3 (Fiore and Rosenblatt, 2023: 771). According to Fiore and Rosenblatt, a K_3-theorist, like Saul Kripke, "is in a position to provide a general philosophical

[12] For sentences σ and ψ with $V(\sigma) = \frac{1}{2}$ and $V(\psi) = 0$, it holds that $V(\sigma \wedge \neg\sigma) = \frac{1}{2}$. Thus, an argument from $\sigma \wedge \neg\sigma$ to ψ is invalid, since while the premise $\sigma \wedge \neg\sigma$ has a designated value, the consequence ψ has no designated value.

explanation to justify her decision to retain some instances of excluded middle and to reject others" (Fiore and Rosenblatt, 2023: 783). A K_3-theorist can endorse LEM for those sentences that are *grounded* (Kripke, 1975: 706), i.e., sentences that are assigned a truth value in the smallest fixed point in the process of interpreting the truth predicate by determining its extension and antiextension (Kripke, 1975: 702 ff.). Ungrounded sentences get a classical truth value because their truth (falsity) is ultimately grounded in some obtaining (or not obtaining) empirical fact. Fiore and Rosenblatt also argue against the view that classical logic has a primary status over non-classical logic, since important metalogical proofs (even the proofs of the classical recapture results) are carried out within the framework of classical logic. They respond to this objection by pointing out that these proofs typically use only *instances* of classical laws, such as LEM. By contrast to the *general* law, these instances are perfectly compatible with a paracomplete logic.

Besides Williamson's abductive defence of a classical treatment of the Liar paradox, there are further arguments in favour of classical logic, which are also significantly based on the criterion of *intertheoretic coherence*. Volker Halbach, for example, employs an abductive argument of this kind to dismiss certain non-classical logics. According to Halbach, a non-classical approach to truth which allows for truth-value gaps or gluts, like K_3 or LP, would have severe ramifications for central epistemic concepts, such as *knowledge*:

> Nonclassical logic is contagious. For instance, if knowledge is defined in terms of, or depends in a certain way on, truth, justified belief, and perhaps other notions, then it is highly likely that a nonclassical treatment of truth will spread to epistemology as well. (Halbach, 2014: 289)

In particular, a gap-theorist would have to accept that a sentence of the form"Leon knows that the sentence on the sheet is true or he does not" lacks truth value, in case the sentence on the sheet is the Liar sentence. As a consequence, a gap-theorist has to accept that a generalization of the form "Leon knows something or he doesn't", and even sentences like "If Leon knows something he believes it", can also turn out to be gappy (see Halbach, 2014: 290). For Halbach, this supposed expansion of non-classical semantics into epistemology comes at a high cost:

> If the proponent of nonclassical logic is prepared to pay this price and to give up classical logic not only for truth but for many other areas as well, then he will face the enormous task of rewriting large parts of philosophy and other disciplines; truth-value gaps and gluts will crop up everywhere [...] (Halbach, 2014: 291).

Thus, for Halbach, one reason to prefer classical logic to non-classical logic is that classical logic fares better with regard to the abductive criterion of *intertheoretic coherence*, namely its better coherence with certain areas of epistemology. Solving the Liar paradox in terms of a non-classical semantics that allows for truth-value gaps or gluts wouldn't cohere with how we analyse and understand important epistemological concepts. Furthermore, adopting a non-classical logic in the realm of epistemology wouldn't satisfy the abductive criterion of an *appropriate balance between strength and simplicity*, since truth-value gaps or gluts would make epistemology unnecessarily complicated.

Halbach's abductive argument in favour of classical logic doesn't seem to be entirely convincing. First of all, proponents of gap-theories or paraconsistent logics accept classical reasoning as long as no gappy or glutty sentences are involved. So, in most cases truth-value gaps and gluts won't crop up. However, if we are explicitly dealing with contexts where paradoxical sentences are involved, it doesn't seem to be counterintuitive that instances of classical logical truths, which apply epistemological concepts to those paradoxical sentences, might no longer be valid. Secondly, a gap-theorist usually accepts a *classical* concept of knowledge. A gap-theorist subscribes, for example, to the principle that one can only know that p, if p is true, where "true" means to be *classically true*. Thus, if p is *neither true nor false*, p can just never be known. *Dialetheists*, like Priest, who believes that the Liar sentences λ expresses a *true contradiction* (a so-called *dialetheia*), seems to accept a traditional analysis of knowledge as well. However, for dialetheists, "true" also includes dialetheias. So, a dialetheist who *believes* that the Liar sentence λ expresses a dialetheia and is *justified* in believing that λ expresses a dialetheia would also claim to *know* that λ expresses a dialetheia, i.e., that λ is a *true* (and, of course, also false) sentence. So, a non-classical semantics need not result in a non-classical epistemology. Gap-theorists and dialetheists have different epistemic attitudes towards the Liar sentence, but not because they adopt different non-classical conceptions of knowledge. Both gap-theorists and dialetheists subscribe to the same (traditional) conception of knowledge, according to which knowledge consists in a true and justified belief (plus possibly an appropriate anti-Gettier clause). It is only their different semantic evaluation of the Liar sentence that explains their divergent epistemic attitudes towards the Liar sentence. In any case, there is no need for gap-theorists and dialetheists to rewrite large parts of epistemology.

So far it has been argued that Williamson's abductive argument in favour of classical logic due to its alleged better intertheoretic coherence with scientific reasoning and its better overall performance with regard to simplicity, expressive strength and explanatory power is controversial. Furthermore, it was argued that Halbach's abductive defence of classical logic, according to which non-classical approaches to the Liar paradox would spread to epistemology, where they give rise to unnecessary complications, does not seem to be entirely compelling.

5 An Abductive Assessment of K_3 and LP and the Revenge Problem

Irrespective of Williamson's and Halbach's objections against K_3 and LP, there could be *other* abductive arguments that might speak against one or both of these non-classical logics as suitable strategies for dealing with the Liar paradox. In the following, the abductive strength of gap- and glut-theories will be analyzed using the abductive criteria introduced earlier.

With regard to *criterion* 2, it is important to examine what evidence and data the two logical theories K_3 and LP intend to address. It is clearly an abductive advantage of both non-classical theories K_3 and LP that they save an unrestricted T-schema and are thus able to accommodate a fundamental intuition about the notion of truth. Gap-theorists, like Kripke, regard paradoxical sentences, such as the Liar sentence λ, to be *ungrounded* (Kripke, 1975: 694). λ's truth value cannot be determined by referring to certain obtaining or non-obtaining states of affairs. Thus, λ lacks a classical truth-value. Nevertheless, we do encounter paradoxical sentences in our language and thought. Some sentences wear their paradoxical character on their sleeves, like the Liar sentence. Other sentences only prove to be paradoxical under certain unfavorable circumstances. For example, the sentence uttered by speaker S_1 "The next sentence that speaker S_2 will utter is true" is only paradoxical in certain contexts, for example, if the next sentence uttered by speaker S_2 is the sentence "What speaker S_1 has just said is false." That is why Kripke, as already mentioned above, requires from an appropriate logical theory of truth that "it must allow our statements involving the notion of truth to be *risky*: they risk being paradoxical if the empirical facts are extremely (and unexpectedly) unfavorable" (Kripke, 1975: 692). So, the idea of paradoxical sentences to be *gappy*, and the phenomenon of certain sentences being *contingently paradoxical*, are important data that a gap-theorist intends to address.

In contrast to gap-theorists, Priest contends that the Liar sentence expresses a *true contradiction*, a *dialetheia*. According to Priest's *dialetheism*, there are things in the world that are actually (or at least possibly) inconsistent. To be sure, contradictions are ubiquitous in our cognitive and scientific activities. People sometimes hold inconsistent beliefs. We sometimes (mostly unintentionally) say things that are inconsistent or imply contradictions. There are also inconsistent obligations and rules. Moreover, some theories in the natural science, mathematics, logic, law, philosophy, etc. contain contradictions. But usually all those contradictions are regarded as some kind of intellectual shortcomings. We feel being mistaken if we encounter such contradictions in our thinking—and thus attempt to resolve them. According to a dialetheist, the contradiction revealed, in particular, by the Liar sentence, is of a different nature: it is not an intellectual shortcoming, but a *true* contradiction, a contradiction made true by an irrevocable fact of the matter

Priest's dialetheism is a *metaphysical view*. Dialetheias exist independently of how we think, reason or talk about things. Their inconsistent nature is not a mere creation of thought, reasoning or language use. Priest's *metaphysical dialetheism* even extends the existence of dialetheias to the empirical world. For example, Priest contends that dialetheias occur in *transition states* when bodies change their place in time (see Priest, 2006: 159–162). In contrast to Priest's metaphysical dialetheism, advocates of *semantic versions* of dialetheism regard true contradictions as purely semantic in character. According to *semantic dialetheism*, dialetheias "arise because of the relationship between language and the world", for example, because of "overdefined" predicates in the natural language (Mares, 2004: 266). The abductive cost-benefit analysis of paraconsistent logics certainly depends on whether contradictions are to be understood merely as inconsistent beliefs or linguistic fallacies that should be avoided or whether they represent indispensable facts of semantics or even of the empirical world. Metaphysical dialetheism is certainly a controversial philosophical position. However, if there were good reasons for assuming the actual existence of inconsistent things in the world, then these dialetheias (in accordance with abductive criterion 2) would be data that a logical theory should consider—and would give a paraconsistent logic, such as LP, an additional abductive weight.

Gap-theorists and glut-theorists both agree on maintaining an unrestricted T-schema which captures an important minimal intuition about sentences with a truth predicate. However, they have different views about contradictions and the semantic status of the Liar sentence. As a result, they intend to account for and focus on different evidence and data, especially with respect to paradoxical sentences. It seems that both logics K_3 and LP are in good agreement with this evidence and data they aim to address.

With regard to abductive *criteria* 3 and 4, one of the often-mentioned disadvantages of K_3 and LP is their apparent vulnerability to *revenge* problems. In K_3, *falsity* is to be distinguished from *untruth*. A sentence σ which is untrue, need not be false since there is the third option of a sentence being *gappy*. A sentence σ being *untrue* in K_3 means, intuitively, that σ is *either false or gappy*. However, this notion of untruth gives rise to a *Strengthened Liar sentence*, a sentence "saying" of itself that it is *untrue*. This sentence appears to be true if and only if it is untrue—which is a contradiction in K_3 and thus leads to K_3's triviality. So, on pain of triviality, the central meta-theoretic notion of *untruth* cannot be part of K_3's object language. Since solving the semantic paradoxes is among K_3's main aims and since the notion of untruth plays a crucial part in achieving this aim, this means a severe limitation of K_3's expressive adequacy.[13]

In LP, truth and untruth are exhaustive and not exclusive. A sentence claiming of itself to be untrue is just both true and untrue. But since ECQ is invalid, triviality doesn't follow. However, it has been objected that on pain of triviality, the notions of *true only* (i.e., *having just the value* 1 *in the model*) or *false only* (i.e., *having just the value 0 in the model*) cannot adequately be expressed *within* a language of LP (see, for example, Bromand, 2002 or Littmann and Simmons, 2004). For a sentence λ^* claiming of itself *not to be true only* (i.e., not to have the value 1 in the model) or a sentence λ^{**} claiming of itself to be *false only* (i.e., to just have the value 0 in the model), there are no consistent models: $V(\lambda^*) = 1$ iff $V(\lambda^*) \neq 1$, and $V(\lambda^{**}) = 1$ iff $V(\lambda^{**}) \neq 1$ for all LP-models—and thus even in LP triviality would follow. It therefore seems that the central semantic notions of *true only* and *false only* cannot adequately be expressed within LP. So, just as K_3, LP also suffers from a severe limitation of expressive power, or so it seems.

Whether this alleged revenge problem is a serious problem for LP is, however, controversial. It could be objected that the definitions of "true only" and "false only" which give rise to an inconsistent model theory for LP, are simply wrongheaded since they are not in line with a full-blooded paraconsistent account. The definitions

[13] Kripke himself was well aware of this revenge problem when famously claiming: "The necessity to ascend to a metalanguage may be one of the weaknesses of the present theory. The ghost of the Tarski hierarchy is still with us" (Kripke, 1975: 714).

of the notions of "true only" and "false only" in terms of "just having value 1 in the model" and "just having value 0 in the model" are, from a paraconsistent point of view, illegitimate attempts to reintroduce the classical exclusive notions of truth and falsity into LP by the backdoor. A paraconsistent logician, in order to avoid triviality due to the reappearance of paradoxes using classical notions of truth and falsity, could therefore consequently apply paraconsistent thinking also to the model-theoretic level. A way of doing this could consist in endorsing a kind of *plurivalent logic*, i.e., a theory which allows sentences to take more than one value—without thereby inducing triviality (for a plurivalent logic, see, for example, Priest, 2018: 89ff.). A sentence saying of itself that it is not true only (or false only) could then be conceived of as a sentence being true only and also not true only (or false only and also not false only). So, it seems that a paraconsistent logic has the resources to account for the problem of a potential loss of expressive adequacy due to possible revenge paradoxes.

Another alleged problem for LP's expressive adequacy is the fact that a Strengthened Liar sentence λ^S, i.e., a sentence claiming of itself to be *untrue*, is, in a sense, "more inconsistent" (see Priest, 2016: 72) than the original Liar sentence λ saying of itself to be false. In LP, λ^S turns out to be not just *true-and-false*, but also *true and untrue*: $Tr\langle\lambda^S\rangle \wedge Fl\langle\lambda^S\rangle \wedge \neg Tr\langle\lambda^S\rangle$, where "$Fl\langle\sigma\rangle$" stands for the falsity of a sentence σ, i.e. for $Tr\langle\neg\sigma\rangle$. So, in particular, we have $(Tr\langle\lambda^S\rangle \wedge Fl\langle\lambda^S\rangle) \wedge (\neg Tr\langle\lambda^S\rangle \vee \neg Fl\langle\lambda^S\rangle)$, i.e., $(Tr\langle\lambda^S\rangle \wedge Fl\langle\lambda^S\rangle) \wedge \neg(Tr\langle\lambda^S\rangle \wedge Fl\langle\lambda^S\rangle)$. For Priest, λ^S, just as the original Liar sentence λ, is a *dialetheia*, i.e., a *true contradiction*, a sentence for which it is true that it is both true-and-false. The Strengthened Liar sentence λ^S, however, is not just a dialetheia (since $(Tr\langle\lambda^S\rangle \wedge Fl\langle\lambda^S\rangle)$), it is also not a dialetheia (since $\neg(Tr\langle\lambda^S\rangle \wedge Fl\langle\lambda^S\rangle)$). So, a dialetheist cannot *with an exclusive demand for truth* express one of the most important consequences of dialetheism—namely that the Strengthened Liar sentence is a dialetheia. But does this really constitute a problem for the expressive adequacy of a dialetheist account? A dialetheist could respond that in a dialetheist framework, a truthful assertion that a certain sentence is a dialetheia isn't annihilated by the fact that the sentence is also not a dialetheia. Just as the Liar sentence simultaneously features both properties, being true and being false, the Strengthened Liar sentence possesses inconsistent semantic properties, namely the properties of being true-and-false and of being not true-and-false. For a dialetheist, there is a fact of the matter that makes it both true that the Strengthened Liar sentence is true-and false and that the Strengthened Liar is not true-and-false. Thus, in asserting that the Strengthened Liar sentence is true-and-false, the dialetheist expresses something true. And this assertion remains correct and significant even if it would have been also correct to claim that the Strengthened Liar sentence is not true-and-false. So again, it seems to me that a paraconsistent account has the resources to adequately express its semantic concept of truth without risking triviality.

In addition, and related to the revenge problem, it has been objected that non-classical approaches, such as paracomplete and paraconsistent accounts, were not capable of a *uniform solution to semantic paradoxes*, since, apart from *truth*, other central semantic concepts could not be expressed within these accounts on pain of triviality. With regard to both *criterion* 3 and *criterion* 4, this would, of course, be an abductive weakening of K_3 and LP. For example, JC Beall and Julien Murzi have argued that adding a *validity* predicate to these non-classical logics would lead to triviality, as a certain *validity* version of Curry's paradox shows (Beall and Murzi, 2013). However, strategies have been developed to prevent a "validity paradox", in particular, within the semantics of LP (see Palios, 2020.).[14] The question of the extent to which paracomplete and paraconsistent approaches are able to provide a satisfactory uniform treatment of different kinds of semantic paradoxes, cannot be analysed within the scope of this paper. However, these discussions show that besides maintaining an unrestricted T-scheme while avoiding triviality, there are further criteria, such as the *expressive power* and *coherence with semantic concepts related to truth*, that play an important role in the abductive methodology which guides the choice of a suitable logic for solving the Liar paradox.

How K_3 or LP are to be evaluated in an overall abductive cost-benefit calculation will, of course, also depend on how alternative approaches perform in relation to the above-mentioned abductive criteria. Alternative *classical* approaches to the semantic paradoxes are numerous and enormously diverse (see Rossi, forthcoming). They are motivated by different formal, semantic and philosophical considerations.

Hierarchical approaches, such as Alfred Tarski's solution to the Liar paradox (Tarski, 1936), do not seem to perfectly address linguistic or semantic intuitions about the notion of truth. In restricting the T-schema in some way or another the resulting theory loses a significant amount of expressive power. In particular, a general truth (and falsity) predicate applied to *all* sentences of a language of a theory T, as used in "All (well-formed) sentences are either true or false," cannot be part of that language itself. Furthermore, as Saul Kripke has famously pointed out, there is no "*intrinsic* criterion that will enable us to sieve out—as meaningless, or ill-formed—those sentences which lead to paradox" (Kripke, 1975: 692). Sentences under unfavourable circumstances can turn out to be paradoxical. So, following Kripke's moral that "an adequate theory must allow our statements involving the notion of truth to be *risky*" (Kripke, 1975: 692), one better opts for a solution to the semantic paradoxes in which sentences that possibly imply paradoxes are not "sieved out" as ill-formed sentences.

Non-hierarchical or *type-free classical approaches* in which the Liar sentence is conceived as a well-formed sentence also bear abductive costs. For example, the proof-theoretic theory FS (based on Friedman and Sheard, 1987) rejects a fully transparent truth predicate in restricting the inference rules *capture* and *release* to those sentences *provable* in FS. The Liar sentence λ, although well-formed, is neither provable in FS nor is its negation. But, of course, since LEM holds, it is provable in FS that $\lambda \vee \neg\lambda$. So, λ is either true or its negation, but FS cannot prove which of both alternatives is the case. One disadvantage of FS is that this logic turns out to be ω-inconsistent (see McGee, 1985, but see Halbach, 1994 for showing that FS is still

[14] See also (Murzi and Rossi, 2020) for new revenge problems for non-classical approaches to the Liar paradox, such as K_3 and LP, using the predicate "is paradoxical"—and a critical assessment of those "paradoxicality paradoxes" in (Nava, 2022).

arithmetically sound). Another expressive limitation of FS is its inability to prove certain universal claims of the iteration of truth (see, in particular, Fujimoto and Halbach, 2018: 730). So, FS, besides from heavily restricting the intuitive T-schema, is not free from certain defects which limit its expressive and explanatory adequacy.

Certain *substructural logics*, in particular a *Strict-Tolerant Logic* that rejects a transitive consequence relation but maintain a transparent truth predicate, (see, for example, Cobreros et al., 2013; Ripley, 2013) seem to be abductively advantageous with respect to a uniform solution to semantic paradoxes—and, thus, with respect to certain aspects of criteria 3 and 4. However, it is contentious whether strict tolerant approaches provide a *philosophically* satisfactory treatment of the Liar paradox. In the Strict-Tolerant-Transparent logic (STTT), a logic that is defined on a strong Kleene model and contains a transparent truth predicate, the Liar sentence λ is both a *theorem* (λ is a STTT-consequence of any sentence) and *refutable* (any sentence is a STT-consequence of λ), which seems inadequate in a context of a classical conception of validity (but see Cobreros et al., 2013: 857–859 for interpreting the semantic status of the Liar sentence in STTT).

Of course, there are many more classical approaches, including *contextualist* positions (see, for example, Glanzberg, 2001, 2004). But it seems fair to say that they do not clearly prove to be abductively better than non-classical approaches, such as K_3 and LP, vis-à-vis all of the abductive criteria.[15]

6 An Operator Account of Truth

As we have seen, the inconsistency that the Liar sentence gives rise to, is an immediate consequence of the intuitive T-schema and central logical rules of Q^+. Therefore, this inconsistency does not depend on any further assumptions one could deny without altering logical principles (operational or structural rules) or the core meaning of truth (given by the T-schema). According to Priest, the futile intensive search for a satisfying "consistent theory" that solves the Liar paradox, provides sufficient grounds for assuming the existence of true contradictions (see Priest, 1998: 424). But if there was a non-trivial treatment of the semantic paradoxes which holds on to the T-schema, but does not analyse the Liar paradox as an inner-logical contradiction, such an account could have abductive advantages. In the following, I will suggest such an account—an account which treats truth as an *operator*.

The way we ascribe truth (or falsity) in natural language does not obviously favour a predicate use. Clearly, we often use "true" in a predicative way when ascribing truth to a name of a sentence, as in "'2 +2 = 4' is true" or "Philitas of Cos' last sentence is false." However, there are many other uses of "true" that suggest an *operational* reading. For example, when claiming "It is true that Yoga makes her happy," we apparently use "true" as a *sentential operator*, i.e., as an operator which

[15] For a discussion of further objections to non-classical approaches to the Liar paradox see (Brendel, forthcoming).

when attached to a sentence (or any other truth bearer, such as a proposition) yields another sentence. So, the heterogeneous grammatical structures of truth ascriptions in natural language do not clearly point in favour of a predicative use of "true".[16]

Treating truth as an operator could be abductively beneficial. In particular, an operator account of truth allows for a certain *intertheoretic coherence with modalities*. Many modal notions whose linguistic and logical structure appear to be very similar to the linguistic and logical structure of the notion of truth, such as "necessary", "possible", "probable", "believe" or "know", are predominately treated as *sentential operators*. So, whereas *truth* is typically formalized as a *predicate*, the *modalities of truth*—such as possibility and necessity—are mostly formalized as *operators*. Thus, the suggestion to formally analyse truth as an *operator*, reduces this asymmetric treatment between truth and modalities.[17]

To examine how an operator approach to truth fares with regard to handling the Liar paradox, let us consider $Q^{+\circ}$ which is, just as Q^{+}, a theory of Robinson arithmetic, but contains (instead of a truth predicate) a sentential operator "T" such that "$T(\sigma)$" is a sentence of the language of $Q^{+\circ}$, expressing, intuitively, that *it is true that* σ (for any sentence σ of $Q^{+\circ}$'s language). Instead of the original T-schema TS, $Q^{+\circ}$ possesses the following operational variant of the T-schema TS°:

$$\text{TS}^\circ :\vdash_{Q^\circ} T(\sigma) \leftrightarrow \sigma \quad \text{(for all sentences } \sigma).$$

In $Q^{+\circ}$ we then have the following inference rules capture° and release°:

capture°: $\sigma \vdash_{Q^{+\circ}} Tr\langle\sigma\rangle$
release°: $Tr\langle\sigma\rangle \vdash_{Q^{+\circ}} \sigma$
(for all sentences σ of the language of $Q^{+\circ}$).[18]

Since "T" is not a predicate, the diagonal lemma cannot be applied—and no semantic paradox gets off the ground. To be sure, a sentence being equivalent to a

---

[16] Among the philosophers who argue for an operational reading of "is true" are, for example, Bobzien (2017), Künne (2003) and Mulligan (2010).

[17] It seems that for the majority of philosophers and logicians, the threat of paradoxes (see, in particular, Montague's paradoxes in Montague, 1963) was a motivating reason to refrain from treating modalities as predicates and instead conceive them as sentential operators. However, the threat of semantic paradoxes didn't provide a reason for most philosophers and logicians to formalize *truth* as an operator. Stern contends that another motivation for an operational treatment of modalities is due to "the impressive story of mathematical success of modal operator logic which originated with the development of possible world semantics by Kanger, Kripke, Hintikka and others" (Stern, 2014a: 552). Whereas Stern attempts to remove the asymmetric treatment between truth and modalities via a *predicate approach to modality* (see Stern, 2014a, 2016), this paper suggests to take the opposite path and vote for an *operator approach to truth*.

[18] So, in $Q^{+\circ}$ the truth operator T is extensionally equivalent with *double negation*. That T functions in $Q^{+\circ}$ just like $\neg\neg$, does, however, not call for criticism. Of course, there is more to truth than double negation. But neither the original T-schema TS nor TS° intends to provide a *definition* of truth. Both schemas are mere minimal principles that any adequate account of truth has to satisfy. Since T is extensionally equivalent to double negation, Q^{+} minus TS is proof-theoretically equivalent to $Q^{+\circ}$. Therefore, the addition of the truth operator T and TS° alone cannot introduce an inconsistency into $Q^{+\circ}$.

sentence claiming that this sentence is false *does* exist in $Q^{+\circ}$. This is a sentence of the form $\sigma \leftrightarrow \neg T(\sigma)$. But this equivalence is, of course, *not provable* in $Q^{+\circ}$. Whereas the existence of a self-referential Liar sentence λ and its provable equivalence with $\neg Tr\langle\lambda\rangle$ via diagonalization lead (together with TS) to a provable contradiction and thus to Q^{+}'s triviality, $Q^{+\circ}$'s sentence of the form $\sigma \leftrightarrow \neg T(\sigma)$ just turns out to be (given TS°) *necessarily false*: $\vdash_{Q^{+\circ}} \neg(\sigma \leftrightarrow \neg T(\sigma))$. So, we have a mere logical falsity instead of a paradox. This avoidance of semantic paradoxes in $Q^{+\circ}$ seems to be gained at the expense of banning self-reference with regard to the notion of truth or falsity—and thus at the expense of $Q^{+\circ}$'s expressive adequacy.

However, under certain circumstances, also sentential operators (as, in particular, *modal* operators) can produce paradoxes (see, for example, Prior, 1961). As we know from Epimenides' paradox, the claim of the Cretan Epimenides that all Cretans are liars, only generates a paradox under very specific, and normally quite unrealistic, assumptions. Let χ be the sentence, asserted by the Cretan Epimenides, that all Cretans are liars, meaning in this context that whatever a Cretan says, is false:

(χ) For all sentences σ : If a Cretan asserts σ, then σ is false.

Let us assume that χ is true. Since χ is a sentence, χ can be used as a substitution instance of itself. And since χ is asserted by the Cretan Epimenides, χ is false. So, by *reductio*, χ must be false, i.e., it is not the case that what all Cretans assert, is false. So, there must be at least one sentence, asserted by a Cretan which is true. In asserting χ, Epimenides thus (probably unintentionally) asserts a necessarily false sentence. Epimenides' assertion only involves a fatal paradox under further assumptions, as the assumption that all the other Cretans (i.e., all Cretans except for Epimenides) lie, and that the only sentence Epimenides ever asserts is χ.

In $Q^{+\circ}$, "Liar-like" paradoxes can be produced in a way similar to Epimenides' paradox. In order to formalise those paradoxes, we have to extend $Q^{+\circ}$ in a way such that $Q^{+\circ}$ allows for *quantification over its own sentences*. Typically, Liar-like sentences λ° have the following general structure:

(λ°) $\forall\sigma(A(\sigma) \rightarrow \neg T(\sigma))$ (for all sentences σ of the language $Q^{+\circ}$)

In order for λ° being paradoxical, further assumptions have to be made such that the following two claims hold:

(1) $A(\lambda^\circ)$
(2) $\neg\exists\sigma(\sigma \neq \lambda^\circ \wedge A(\sigma) \wedge T(\sigma))$ (i.e. there is no sentence different from λ° for which λ° is true).

It is easy to see that a sentence of the form λ° together with (1) and (2) lead to a contradiction in $Q^{+\circ}$: if we assume that λ° is true, i.e., T(λ°), we get, with release°, λ°—and thus $\forall\sigma(A(\sigma) \rightarrow \neg T(\sigma))$. With universal instantiation, we then get $A(\lambda^\circ) \rightarrow \neg T(\lambda^\circ))$, and, therefore, with (1): $\neg T(\lambda^\circ)$. So, by *reductio*, we get $\neg T(\lambda^\circ)$—and therefore, via capture°—$\neg\lambda^\circ$, i.e., $\exists\sigma(A(\sigma) \wedge T(\sigma))$. Let σ^* be such a σ for which $A(\sigma) \wedge T(\sigma)$. i.e., $A(\sigma^*) \wedge T(\sigma^*)$. From the latter, we get with (2) : $\sigma^* = \lambda^\circ$, and thus $T(\lambda^\circ)$. So, we have deduced a contradiction which (via ECQ) leads to trivialisation.

For paradoxes gained via a truth operator, the above sentence schema $A(\lambda^\circ)$ usually expresses a certain discursive, cognitive, or epistemic relation (such as speaking, saying, asserting, uttering, thinking, believing etc.) with regard to sentence λ°, and (2) expresses a given circumstance under which λ° turns out to be paradoxical. So, the paradox arises because someone asserts (utters, thinks, believes etc.) a sentence, and in asserting (uttering, thinking, believing etc.) this sentence, one thereby happens to produce an inconsistency in the given situation. It is important to notice that in deriving the inconsistency (and triviality) of the given logical system with a truth operator, we utilised the language's capability of *self-application*. Instead of forming a *self-referential* sentence (such as the Liar sentence λ in Q^+), we started from a universally quantified sentence λ° to which λ° was self-applied as its own true substitution instance. So, instead of self-reference, self-application is the potential source of inconsistency in $Q^{+\circ}$. But whereas *self-reference* is an *inner-logical* feature due to the diagonal lemma in Q^+, *self-application* in $Q^{+\circ}$ is only possible when certain *extra-logical* assumptions are made.

In order to apply the above general schema of producing a paradox in a logical theory with a truth operator to Epimenides' paradox, let C be an operator in $Q^{+\circ}$ such that $C(\sigma)$ reads as "*σ is asserted by a Cretan.*" The sentence χ, asserted by Epimenides, is then the sentence:

$(\chi)\ \forall\sigma(C(\sigma) \rightarrow \neg T(\sigma))$ for all sentences σ of the language of $Q^{+\circ}$.

Since Epimenides is a Cretan, it holds that:

(1) $C(\chi)$

We furthermore assume that no other sentence asserted by a Cretan, except from the sentence χ that Epimenides has asserted, is true, i.e.:

(2) $\neg\exists\sigma(\sigma \neq \chi \wedge C(\sigma) \wedge T(\sigma))$

As seen above, σ generates a paradox under the assumption of (1) and (2).

To be sure, in Epimenides' paradox, assumption (2) is highly unrealistic. There are, however, other more realistic versions of those kinds of paradoxes produced within a logical theory using a truth operator. Here is one example: let $Th^S_\circ$ be an additional operator in $Q^{+\circ}$ which reads as "Subject S is thinking at the moment t_0 that". We can now formalize the sentence (abbreviated by τ) saying that everything S is thinking at t_0 is false in the following way:

$(\tau)\ \forall\sigma(Th^S_\circ(\sigma) \rightarrow \neg T(\sigma))$ (for all sentences σ of the language of $Q^{+\circ}$).

Let us abbreviate this sentence by τ. We now assume that the subject S is thinking at t_0 that τ, and that this is the *only thought* S has at t_0, i.e.:

(1) $Th^S_\circ(\tau)$
(2) $\neg\exists\sigma(Th^S_\circ(\sigma) \wedge \sigma \neq \tau)$.[19]

---

[19] Since τ is the only thought of S at t_0, we can omit the conjunct $T(\sigma)$.

Again, sentence τ together with (1) and (2) lead to a contradiction in $Q^{+\circ}$—and thus to triviality.

The above produced paradoxes show that treating truth as a sentential operator does not provide a universal remedy against all kinds of semantic paradoxes. An extended logical theory $Q^{+\circ}$ which allows for sentential quantification and contains some further discursive, cognitive or epistemic sentential operators, leads, under certain circumstances, to inconsistency and triviality. That is why Johannes Stern claims: "If the aim is to provide an adequate account of truth and the modalities within natural language, especially English, any account, no matter whether it treats truth and modalities as predicates or operator, will have to face the paradoxes at some stage" (Stern, 2014b: 77).[20]

However, as the above schematic analysis of paradoxes with a truth operator shows, there is a difference between truth operator paradoxes and the Liar paradox in predicative accounts of truth. In contrast to the Liar paradox in $Q+$ in which a contradiction was derived via the *inner-logical* diagonal lemma together with the T-schema, the paradoxical result in $Q^{+\circ}$ can only be gained from TS° with the help of the *extra-logical assumptions* (1) and (2). Thus, in a predicative account of truth, truth is, in a certain sense, *intrinsically* paradoxical. Whereas a truth *operator*, minimally characterized by TS°, does not by itself create a paradox in $Q^{+\circ}$. According to an operator approach, truth is therefore not intrinsically paradoxical. Certain ascriptions of truth or falsity imply a contradiction only under certain contingent circumstances.[21]

---

[20] The use of sentential quantification could be seen as an additional abductive cost, since this type of quantification exceeds the means of classical first-order logic. However, sentential quantifications appear in natural language and cannot always be replaced by first-order quantifications, i.e., by quantifiers binding variables standing in singular term position. They are especially useful in contexts of propositional modal logic and allow a universal, non-hybrid, way of quantifying over sentences that combine truth with modality—for a discussion of possible abductive advantages of sentential (propositional) quantifiers over first-order quantifiers, see (Fritz, 2024: 6f).

[21] It could be objected that even in $Q^{+\circ}$ extended by sentential quantification contradictions arise, for example, through diagonalization on the formula $\forall\sigma((\langle\sigma\rangle = x) \rightarrow \neg T(\sigma))$. However, the universal quantifier in this formula only binds the propositional variable σ in the argument position of the negated truth operator, but not the name of σ, since $\langle\sigma\rangle$ is an *individual constant*. A paradox would only arise if the sentential quantifier were interpreted in a certain *substitutional* way that admits quantification into quotation (see, for example, Dunn and Belnap, 1968). As it is well known, quantification into quotation can generate paradox, no matter whether "true" is taken to be a predicate or an operator: by diagonalization on $\forall\sigma((\langle\sigma\rangle = x) \rightarrow \neg T(\sigma))$, there is a sentence λ, such that $\lambda \leftrightarrow \forall\sigma((\langle\sigma\rangle = \langle\lambda\rangle) \rightarrow \neg T(\sigma))$. Assume λ, i.e., $\forall\sigma((\langle\sigma\rangle = \langle\lambda\rangle) \rightarrow \neg T(\sigma))$. If the sentential universal quantifier allowed for quantification into quotation, we would get $((\langle\lambda\rangle = \langle\lambda\rangle) \rightarrow \neg T(\lambda)$ with universal instantiation—and, thus, $\neg T(\lambda)$. From TS°, $\neg\lambda$ would follow. From $\neg\lambda$, however, $\exists\sigma((\langle\sigma\rangle = \langle\lambda\rangle) \wedge T(\sigma))$ follows. Again, if the sentential existential quantifier were interpreted in a way that allows for quantification into quotation, we would get (with existential elimination) $((\langle\sigma^*\rangle = \langle\lambda\rangle) \wedge T(\sigma^*)$—and, therefore, T($\lambda$), i.e. λ. So, we have derived the contradiction $\lambda \leftrightarrow \neg\lambda$. With a truth *predicate*, we could prove the same contradiction (under the assumption that sentential quantifiers are interpreted in a substitutional way such that quantification into quotation is possible) by diagonalization on $\forall\sigma((\langle\sigma\rangle = x) \rightarrow \neg T\langle\sigma\rangle))$. It is, therefore, important to note that the sentential quantifiers do not bind any sentences inside quotation marks. "σ" within "$\langle\sigma\rangle$" is part of a constant singular term, which, like other individual constants, does not fall within the

Treating truth as an operator instead of a predicate could have abductive advantages. With a truth operator no self-referential semantic paradoxes can be produced. However, this does neither mean a limitation of the *adequacy with evidence* nor a lack of *expressive* and *explanatory strength.* As we have seen, an operator account of truth allows to formulate "Liar-like" sentences λ° in form of a universal generalization of which λ° itself is a true instantiation. Only if we allow for an *unrestricted sentential quantification* and introduce a further sentential operator substituted for "A" such that the assumptions (1) and (2) hold, a paradox gets off the ground. So, in particular, without the assumption (1) and (2) which are *external* to $Q^{+\circ}$, TS° remains consistent and, therefore, non-trivial. If, as for example in mathematical uses of logic, we are not interested in the self-applying feature of language and reasoning, the consistency (and non-triviality) of a classical logic with a truth operator can be maintained despite of an unrestricted T-scheme (TS°).

However, if a logical system should also address self-applicational features of language and thought, one has to make further extra-logical assumptions, such as (1) and (2) and allow for an unrestricted sentential quantifier. With such further assumptions a classical logical theory $Q^{+\circ}$ would give rise to a contradiction and to triviality. In such a situation, for example, a non-classical *paraconsistent* approach seems to be a plausible way to avoid the consequence of trivialization. But without the possibility of sentential quantification and extra-logical assumption, such as (1) and (2), $Q^{+\circ}$ is protected against the contradictory consequences that arise from Liar sentences.

The view that paradoxes of self-application only arise through sentential quantification under the additional assumption of non-logical principles does not, however, seem to suggest a *metaphysical* interpretation of dialetheism, according to which true contradictions actually exist in the world. Paradoxes of semantic self-application are possible due to the great flexibility of human language and thought. If one becomes aware that a statement or train of thought leads to paradoxical consequences of semantic self-application, such as Epimenides' statement χ or the sentence (τ), this statement or train of thought certainly does not lead to the trivialisation of speech or thought. Nevertheless, one is inclined to reject these paradoxical consequences and to correct them accordingly—but not to accept these paradoxes as actual facts in the world.

---

scope of the sentential quantifiers in $Q^{+\circ}$. The fact that sentential quantification into quotation is not possible could be seen as an additional *expressive restriction* of $Q^{+\circ}$ (and also of predicative truth approaches with sentential quantifiers). In many *axiomatic* approaches, quantifying over *codes* of sentences is possible, such that, in particular, general claims about the relationship between validity and truth within the object language can be formulated. However, in these approaches the T-schema is restricted, which can also be understood as a certain expressive limitation (see FN 23.).

7 Conclusion

The method of logical abductivism was introduced as a certain kind of logical anti-exceptionalism in the epistemology of logic. There are, in particular, four important criteria that guide an abductive methodology in the assessment of logical theories: *non-triviality*, *adequacy with evidence or data*, *appropriate balance between strength and simplicity*, and *intertheoretic coherence*. These criteria were employed in order to critically analyse abductive arguments in favour of classical approaches to the Liar paradox instead of non-classical solutions, such as K_3 or LP. It was argued that neither Williamson's nor Halbach's abductive reasons for preferring classical logic to gap- or glut-theories are completely convincing. Moreover, many classical solutions to the Liar paradox do not appear to fare particularly better than K_3 or LP with regard to the proposed abductive criteria. However, LP seems to be abductively advantageous over K_3, at least with regard to LP's better expressive adequacy due to LP's superior way to deal with certain potential revenge problems. In the last part of the paper, an *operator account of truth* was suggested as an abductively justified way to handle the Liar paradox. Although there seems to be linguistic evidence that the notion of truth can be used as a sentential operator in natural language, operator accounts of truth have been (with a few exceptions) mostly neglected in the discussion about treatments of the Liar paradox. Besides its evidential adequacy, an operator account of truth has abductive benefits. One positive side-effect of this account is that truth and modalities (which are mainly conceived as operators) are logically treated symmetrically. Furthermore, in an operator account, paradoxicality is no longer an intrinsic property of the concept of truth, but rather a contingent consequence due to features of self-application of language or thought.[22]

In view of the vast diversity of both classical and non-classical solutions to the Liar paradox, it is not the aim of this paper to argue for $Q^{+\circ}$ as the overall abductively

[22] There is an enormous range of both classical and non-classical approaches with a *predicative* account of truth. It is, therefore, not possible to fairly compare the truth operator account $Q^{+\circ}$ with this multitude of solution strategies to the Liar paradox in relation to all of the proposed abductive criteria. On the side of *classical* approaches, only hierarchical approaches, the theory FS and substructural solutions were mentioned and briefly abductively assessed in this paper. But there are also consistent classical approaches that allow for self-reference in which a language of first-order arithmetic is extended with axiom schemas and rules that govern a validity and truth *predicate*, as, for example the account proposed by Ketland (2012). Although self-referential sentences using a validity or truth predicate can be derived via diagonalization, inconsistency is blocked in Ketland's account, since the derivation of such self-referential sentences (for example, a sentence "saying" of itself to be not valid) must use only axioms or rules of *logic*. I do not intend to claim that Ketland's and other classical axiomatic approaches have abductive disadvantages compared to truth operational approaches in *every* respect. Much depends, of course, on what data and (mathematical or philosophical) applications a theory aims to address. The axiomatic approaches do not use an unrestricted T-schema, but introduce truth in the form of axioms weaker than TS. Furthermore, in Ketland's account, $Val(\langle\sigma\rangle)$ can only be inferred for those sentences σ for which there is a logical derivation (where 'Val' stands for the validity predicate), and the truth predicate is defined in terms of sentences of the form $Val(x)$. For applications outside of logical theorems, this appears to be a restriction of the expressive power of truth. In *this* sense, one could argue that $Q^{+\circ}$ might be more abductively advantageous.

best solution. This paper intends to take the Liar paradox as an example to show how abductive criteria (which can similarly be found in scientific theory choice) help to rationally evaluate various approaches to the Liar paradox. Furthermore, this paper aims to point out a previously rather underexplored solution to the Liar paradox in the form of an *operator account of truth*. It was argued that this account seems to meet important abductive criteria successfully. Which approach is finally preferred in an abductive balancing of the various criteria, depends on many factors, including the intended area of application of the logical theory.

Acknowledgements This work was supported by the DFG (Deutsche Forschungsgemeinschaft), research unit FOR 2495. The author is grateful to two anonymous reviewers for their helpful comments on an earlier draft of this paper.

References

Ayer, A. (1952). *Language, truth and logic*. Dover Publications.

Bealer, G. (1998). Intuition and the autonomy of philosophy. In M. R. DePaul & W. Ramsey (Eds.), *Rethinking intuition* (pp. 201–239). Rowman & Littlefield.

Beall, J. C., & Murzi, J. (2013). Two flavors of Curry's paradox. *The Journal of Philosophy, 110*, 143–165.

Bobzien, S. (2017). Gestalt shifts in the liar or why KT4M is the logic of semantic modalities. In B. Armour-Garb (Ed.), *Reflections on the liar* (pp. 71–113). Oxford University Press.

Boolos, G., Burgess, J. P., & Jeffrey, R. C. (2002). *Computability and logic*. Cambridge University Press.

Brendel, E. (forthcoming). Non-classical approaches to semantic paradoxes. In F. Ferrari, E. Brendel, M. Carrara, O. Hjortland, G. Sagi, & G. Sher (Eds.), *Oxford handbook of philosophy of logic*. Oxford University Press.

Bromand, J. (2002). Why paraconsistent logic can only tell half the truth. *Mind, 111*(444), 741–749.

Bueno, O. (2010). Is logic a priori? *The Harvard Review of Philosophy, 17*, 105–117.

Bueno, O., & Colyvan, M. (2004). Logical non-apriorism and the 'law' of non-contradiction. In G. Priest, J. C. Beall, & B. Armour-Garb (Eds.), *The law of non-contradiction* (pp. 156–175). Oxford University Press.

Cobreros, P., Egré, P., Ripley, D., & van Rooij, R. (2013). Reaching transparent truth. *Mind, 122*, 841–866.

da Costa, N., & Arenhart, J. (2018). Full-blooded anti-exceptionalism about logic. *Australasian Journal of Logic, 15*, 362–380.

Dunn, J. M., & Belnap, N. D. (1968). The substitution interpretation of the quantifiers. *Nous, 2*, 177–185.

Fiore, C., & Rosenblatt, L. (2023). Recapture results and classical logic. *Mind, 132*, 762–788.

Frege, G. (1893). In *Basic laws of logic* (P. Ebert, M. Rossberg, & C. Wright Trans. and Eds.). Oxford University Press (2013).

Friedman, H., & Sheard, M. (1987). An axiomatic approach to self-referential truth. *Annals of Pure and Applied Logic, 33*, 1–21.

Fritz, P. (2024). *Propositional quantifiers*. Cambridge University Press.

Fujimoto, K., & Halbach, V. (2018). The axiomatic approach to truth. In M. Glanzberg (Ed.), *Oxford handbook of truth* (pp. 718–738). Oxford University Press.

Glanzberg, M. (2001). The liar in context. *Philosophical Studies, 103*, 217–251.

Glanzberg, M. (2004). A contextual-hierarchical approach to truth and the liar paradox. *Journal of Philosophical Logic, 33*, 27–88.
Halbach, V. (1994). A system of complete and consistent truth. *Notre Dame Journal of Formal Logic, 35*, 311–327.
Halbach, V. (2014). *Axiomatic theories of truth*. Cambridge University Press.
Hjortland, O. (2017). Anti-exceptionalism about logic. *Philosophical Studies, 174*, 631–658.
Hjortland, O. (2019). What counts as evidence for a logical theory? *Australasian Journal of Logic, 16*, 250–282.
Kant, I. (1781). *Critique of pure reason* (P. Guyer & A. W. Wood, Trans.). Cambridge University Press (1998).
Ketland, J. (2012). Validity as a primitive. *Analysis, 72*, 421–430.
Kripke, S. (1975). Outline of a theory of truth. *Journal of Philosophy, 72*, 690–716.
Künne, W. (2003). *Conceptions of truth*. Clarendon Press.
Littmann, G., & Simmons, K. (2004). A critique of dialetheism. In G. Priest, J. C. Beall, & B. Armour-Garb (Eds.), *The law of non-contradiction* (pp. 314–335). Oxford University Press.
Maddy, P. (2002). A naturalistic look at logic. *Proceedings and Addresses of the American Philosophical Association, 76*, 61–90.
Mares, E. (2004). Semantic dialetheism. In G. Priest, J. C. Beall, & B. Armour-Garb (Eds.), *The law of non-contradiction* (pp. 264–275). Oxford University Press.
Martin, B., & Hjortland, O. (2022). Anti-exceptionalism about logic as tradition rejection. *Synthese, 200*, 148.
McGee, V. (1985). How truthlike can a predicate be? *Journal of Philosophical Logic, 14*, 399–410.
Montague, R. (1963). Syntactical treatments of modality, with corollaries on reflexion principles and finite axiomatizability. *Acta Philosophica Fennica, 16*, 153–167.
Mulligan, K. (2010). The truth predicate vs the truth connective: On taking connectives seriously. *Dialectica, 64*, 565–584.
Murzi, J., & Rossi, L. (2020). Generalized revenge. *Australasian Journal of Philosophy, 98*(1), 153–177.
Nava, W. (2022). Dissolving the paradoxicality paradox. *Australasian Journal of Logic, 19*, 133–146.
Palios, F. (2020). Validity, dialetheism and self-reference. *Synthese, 197*, 773–792.
Priest, G. (1979). Logic of paradox. *Journal of Philosophical Logic, 8*, 219–241.
Priest, G. (1998). What is so bad about contradictions? *The Journal of Philosophy, 95*, 410–426.
Priest, G. (2006). *In contradiction* (2nd ed.). Oxford University Press.
Priest, G. (2008). *An introduction to non-classical logics* (2nd ed.). Cambridge University Press.
Priest, G. (2014). Revising logic. In P. Rush (Ed.), *The metaphysics of logic* (pp. 211–223). Cambridge University Press.
Priest, G. (2016). Logical disputes and the a priori. *Princípios: Revista de Filosofia, 23*, 29–57.
Priest, G. (2018). *The fifth corner of four*. Oxford University Press.
Prior, A. (1961). On a family of paradoxes. *Notre Dame Journal of Formal Logic, 2*, 16–32.
Putnam, H. (1969). Is logic empirical? In R. S. Cohen & M. W. Wartofsky (Eds.), *Boston Studies in the Philosophy of Science: Proceedings of the Boston Colloquium for the Philosophy Science 166/1968* (pp. 216–241). Springer.
Ripley, D. (2013). Paradoxes and failures of cut. *Australasian Journal of Philosophy, 91*, 139–164.
Rossi, L. (forthcoming). Classical solutions to semantic paradoxes. In F. Ferrari, E. Brendel, M. Carrara, O. Hjortland, G. Sagi, & G. Sher (Eds.), *Oxford handbook of philosophy of logic*. Oxford University Press.
Russell, G. (2014). Metaphysical analyticity and the epistemology of logic. *Philosophical Studies, 171*, 161–175.
Russell, G. (2019). Deviance and vice: Strength as a theoretical virtue in the epistemology of logic. *Philosophy and Phenomenological Research, 99*, 548–563.
Sainsbury, R. (2002). What logic should we think? *Royal Institute of Philosophy Supplements, 51*, 1–17.

Stern, J. (2014a). Montague's theorem and modal logic. *Erkenntnis,79*, 551–570.
Stern, J. (2014b). The truth on predicates and connectives. In A. Reboul (Ed.), *Mind, values, and metaphysics. philosophical essays in honor of Kevin Mulligan* (Vol. 1, pp. 63–78). Springer.
Stern, J. (2016). *Toward predicate approaches to modality*. Springer.
Tarski, A. (1936). On the concept of logical consequence. In *Logic, semantics, metamathematics* (J. H. Woodger, Transl., 2nd ed., pp. 63–78). Hackett (1983).
Williamson, T. (2013). *Modal logic as metaphysics*. Oxford University Press.
Williamson, T. (2017). Semantic paradoxes and abductive methodology. In B. Armour-Garb (Ed.), *Reflections on the liar* (pp. 325–346). Oxford University Press.
Williamson, T. (2018). *Doing philosophy: From common curiosity to logical reasoning*. Oxford University Press.

Bootstrapping a Logical Theory

Gillman Payette and Nicole Wyatt

Abstract In this paper we consider whether the background logic problem actually presents a serious challenge for anti-exceptionalists about logic. We argue that logical theorizing does not depend upon a background logic to get started, but merely on the defeasible assumption that the specific arguments offered are good ones. These assumptions are compatible with multiple explanations of the goodness of these arguments, including accounts on which they are not formally valid but materially justified. Furthermore, as in other areas of science, the success of the general research program is compatible with the rejection of initial characterizations of the data—in this case initial claims about the goodness of specific arguments—as well as rejections of initial hypotheses about both the logical form of those specific arguments and the explanation of their goodness.

1 The Problem

Anti-exceptionalists maintain that logic is not substantially different from the accepted empirical sciences in either its subject matter or its epistemology. There are, predictably, lots of different variants of anti-exceptionalism. We largely agree with Martin and Hjortland (2022) in viewing anti-exceptionalism as unified by *tradition rejection*, and the variations as dependent on which parts of the tradition you reject.

For us in this paper the key claim is that the epistemology of logic does not involve *a priori* access to either data or putative logical laws or rules. In particular, anti-exceptionalists of this variety frequently advocate for the view that the process of theory choice in logic is similar to that of the sciences.[1] The idea that theory choice in

[1] For example, Maddy (2002), Williamson (2013), Priest (2014), Russell (2015), Hjortland (2017), Hjortland (2020) and Martin (2020).

G. Payette (✉) · N. Wyatt
University of Calgary, Calgary, Alberta, Canada
e-mail: ggpayett@ucalgary.ca

N. Wyatt
e-mail: nicole.wyatt@ucalgary.ca

S. Molick (ed.), *Demarcating Logic and Science*, Studies in Applied Philosophy, Epistemology and Rational Ethics 80, https://doi.org/10.1007/978-3-032-25674-4_3

logic is, epistemologically speaking, mundane is nicely captured by Richard Sylvan (née Routley) as follows:

> Choice of a logical theory is a special case of the choice of a theory or a system, and choice of these does not differ in principle from choice of such diverse items as a new house, a winner (e.g. of a gymnastics or equestrian contest), or of a recording of a symphony. The special features lie only in the factors considered and the constraints adopted. The familiar procedure of laying down conditions of adequacy is tantamount to the working out of constraints. (Routley, 1980, 81)

However, logical theory choice has been thought to be subject to a special epistemological problem not present for theory choice in the sciences, or indeed choice more generally—the *background logic problem*. The worry is this: the process of choosing a theory depends in part upon making various inferences, so it seems that a logical theory which legitimatizes these inferences must already be in place before we can use them to guide our choice of logical theory. A pernicious circularity ensues: the choice of a logical theory depends upon us already having a chosen a logical theory. Moreover, it seems that where we have more than one logical theory between which we are choosing, the choice of a background logic inevitably begs the question against those logical theories in competition with the background logic.[2]

The challenge posed by the background logic problem then is one of justification—the arguments offered in support of a particular choice of logical theory must themselves be justified, but this justification must, it is claimed, presuppose the validity of these very arguments, and thus a logical theory under which these arguments are valid.[3]

Recently, Hjortland and Martin (2023) have argued that avoiding the background logic problem provides some of the best motivation for exceptionalism about logic. If logical theory-choice is justified on some immediate and *a priori* basis, such as logical intuition or grasping the meaning of logical operators, then there is no need for argument to take us from evidence to theory. The anti-exceptionalist on the other hand, takes logical theories to be justified by arguments, thus opening the door for the background logic problem.

More specifically Martin and Hjortland suggest that given any argument for a particular logical theory we are faced with two equally unpalatable options. If the chosen logical theory endorses as valid the argument(s) that led to its adoption, then the argument(s) in question beg the question against those logical theories that do not endorse them as valid. On the other hand, if the chosen logical theory does not endorse the arguments that led to its adoption as valid then it undermines its own justification.

---

[2] Susan Haack presents what might be the earliest version of this problem, describing it as parallel in structure to Hume's problem of induction: "I propose another dilemma: we cannot justify deduction inductively, because to do so would be, at best, to show that usually, when the premisses (sic) of a deductive argument are true, the conclusion is true too—which would be too weak; and we cannot justify deduction deductively, either, because such a justification would be circular" (1976, 112). In Wright (1986) it appears as the *centrality problem*.

[3] C.f. Hjortland and Martin (2023) and Finn (2019).

To sum up, then, the worry for the anti-exceptionalist is that in order to provide *empirical* evidence for a logical theory we will need to make assumptions about logical validity, and this creates a form of justificatory circularity that is not present in other research areas.

2 Responses

Martin and Hjortland describe two options they take to be available for the anti-exceptionalist seeking to address the background logic problem. The first, which they dub the *intra-theoretic approach*, is to simply accept that we cannot start the process of theory choice without presupposing some logical laws. As with the natural sciences, we begin with the starting point we have, and work to revise in a justified manner as new evidence arises (Hjortland and Martin, 2023, 477–8).

The alternative to this—the *inter-theoretic approach*—is to restrict arguments for a particular choice of theories to only those inference rules that are not under dispute—in other words, to those patterns of inference that are agreed upon by all parties (Hjortland and Martin, 2023, 478).

Our view, by way of a preview, is that these two approaches do not exhaust the options, and that the background logic problem rests on assumptions that the anti-exceptionalist can and should reject. However the concerns Martin and Hjortland raise about these two options are useful in identifying these assumptions, and so we will begin with them.

2.1 The Intra-framework Approach

On this approach logical theorizing proceeds from whatever logical framework we have, and then works to revise in a justified manner as new evidence arises. Martin and Hjortland suggest Priest (2014) as an exemplar of this view, and he does seem to have this sort of process in mind when he writes:

> Let me finish this discussion by returning, by way of illustration, to the replacement of traditional logic by mathematical logic in the early years of the tenth century. In the nineteenth century, much new data had turned up: specifically the microscope had been turned on mathematical reasoning, showing all sorts of inferences that did not fit into traditional logic. Mathematical logic was much more adequate to this data. This is not to say that enterprising logicians could not try to stretch traditional logic to account for these inferences. But mathematical logic scored high on many of the other theoretical criteria: simplicity, unifying power, and so on. It was clearly the much better theory. (Priest, 2014, 217)

One objection raised by Martin and Hjortland to the intra-theoretic approach is that this would to allow two research programs to use the same evidence to confirm their own starting logical theories. Both could proceed without ever converging on a single logical theory (Hjortland and Martin, 2023, 478).

There are two assumptions underlying this objection that we wish to question. First, it's not obvious that there is a single true logical theory on which to converge, given the many substantive reasons that have been offered in favour of logical pluralism. Whatever merit there is to the idea that competing research programs should eventually converge on *the* truth depends on there being a singular truth to converge on. It also depends on the theories being *true*. There is a well-established position in the philosophy of science which argues that correct theories may well fall short of truth.[4]

Second, this picture of the scientific process doesn't match all the ways in which competition between research programs actually resolves. The convergence idea may describe the situation in which two labs pursue a different vaccine for an infectious disease. Both these labs are working within the same basic understanding of disease and infection. But consider the competition between the miasma and germ theories of disease transmission. The history of epidemiology is not adequately described as one in which the miasma theorists and the germ theorists started with a single set of evidence and slowly through a process of investigation and argument converged on the same theory. Part of the success of germ theory involved correctly identifying what counted as relevant data—what the evidence was—and what did not. Smells out, infection maps in. Convergence may apply when scientists are attempting to answer a single relatively narrow question within a larger shared explanatory framework, but it seems out of place when the fundamental assumptions surrounding a phenomenon are at issue. The debate between classical and paraconsistent logics is an example of disagreement about the fundamentals.

Let's look at the example of germ vs miasma theories in slightly more detail. There are two ways in which the idealization of competing research programs working with the same evidence eventually converging on the truth falls short. Part of the dispute between the two theories concerned what the relevant data were. Inoculation for small pox was successfully practiced in Europe in the early nineteenth century,[5] but this was not considered relevant by defenders of the miasma view—it was not a piece of data they attempted to explain or accommodate. In 1854, when John Snow produced his map of the cholera infections around the Broad Street pump, he invented a whole new kind of evidence—epidemiological mapping—which provided indirect support for the germ theory.[6] On the other hand, advocates of the germ theory generally ignore any correlation between unpleasant smells and disease transmission, which the miasma theory makes central use of.

In addition, the practical success of germ theory pre-dated empirical confirmation via new instruments or methodology, and indeed, it was the slow adoption of

---

[4] See e.g. Cartwright (1983), Wimsatt (2007), Hindriks (2008), and Woody (2008).

[5] Outside of Europe inoculation was practiced in China as early as the tenth century (Lombard et al, 2007), and medieval Islamic scholars understood that disease could contaminate the water supply Reid (2013).

[6] The evidence is indirect because what Snow's map directly supports is the idea that the Broad Street pump is the source of the infection in the neighbourhood. This is incompatible with the miasma theory, since the pump is not a source of bad odours, but not in and of itself a positive reason to adopt the germ theory.

the theory in late nineteenth century Europe that drove the technological and theoretical developments that eventually provided that confirmation. Germ theory was rendered increasingly plausible because of empirical investigation with the microscope. However this was not enough alone to establish a connection between disease and the 'animacules'—we also needed more sophisticated methodology, such as that provided by Koch's Postulates.[7] Theory choice in science is a mixture of theoretical virtues, one of which is fit to data. But the data are not naively given, as the substantial literature on the theory-ladeness of data in science shows.[8]

Indeed the shift from traditional to mathematical logic is in some important respects quite like the case of germ theory, as Priest himself is aware:

> Given any theory, in science, metaphysics, ethics, logic, or anything else, we choose the theory which best meets those criteria which determine a good theory. Principle amongst these is adequacy to the data for which the theory is meant to account. In the present case, these are those particular inferences that strike us as correct or incorrect. This does not mean that a theory which is good in other respects cannot overturn aberrant data. As is well recognised in the philosophy of science, all things are fallible: both theory and data. (Priest, 2014, 217)

One example of the data overturned by mathematical logic is *Darapti*: the inference from 'All As are Bs' and 'All As are Cs' to 'Some Bs are Cs'. Aristotle took this to be valid for reasons which remain clear, even if we must forcibly set aside our training to see it. It's not that mathematical logic is a better fit with all the data considered by traditional logic. Rather, mathematical logic prioritizes fit with certain kinds of data—mathematical reasoning, primarily—over fit with more ordinary cases, to the point where we no longer see the goodness of *Darapti* patterns of inference as data to be accommodated at all, much as the correlation of bad smells with disease is no longer seen as data for epidemiology. And, just as the epidemiologist can say something about why, historically, people might have noticed an apparent connection between sickness and bad smells, contemporary logicians can address Darapti reasoning as an enthymeme requiring 'A's exist' as a premise.

The second concern raised by Martin and Hjortland for the intra-framework approach is theoretical "flip-flopping". It is in principle possible, they suggest, for there to be two logical theories L_1 and L_2 such that when either is combined with the same data set, it advocates switching to the other logical theory. L_1 plus the data suggests adopting L_2 and vice versa (Hjortland and Martin, 2023, 478).

The underlying model of inquiry presupposed by this objection is one in which we lay out the logical evidence, and then engage in a series of deductive (more on this later) arguments leading to a logical theory. But as we hope the discussion above makes clear, this is not an accurate picture of the general process of theory choice in scientific contexts. The anti-exceptionalist who maintains that theory choice in logic operates in much the same way as it does elsewhere in science has a much broader range of epistemological resources with which to decide between L_1 and L_2. Indeed

---

[7] See Evans (1976) and Nutton (1983) for discussion of the postulates and how they were refined.

[8] Cf. Kuhn (1962) and Feyerabend (1981). Of course this problem is not one that is insurmountable in all cases, see Bogen and Woodward (1992), (Boyd and Bogen, 2025) and Bokulich (2020).

Martin and Hjortland acknowledge that additional methodological principles could offer a resolution to this situation.

Moreover this picture of theory choice as a process of deductive argument does not properly reflect the actual process by which theories, logical and otherwise, are developed and chosen between. We maintain that a more practice-based approach that considers actual practices of theory choice goes a long way towards alleviating this somewhat abstract concern. Indeed our view is that the deductive picture underlying the flip-flopping objection fundamentally mischaracterises the process of logical inquiry, which involves far more tinkering and far less argument.[9]

2.2 *Inter-framework Approach*

Recall that Hjortland and Martin's second proposed approach to the background logic problem is to restrict arguments for a particular choice of theories to only those inference rules that are not under dispute—in other words, to those patterns of inference that are agreed upon by all parties.

This requires, Martin and Hjortland point out, that there be some inference rules that are not under dispute. Once we consider comparing all logical theories at once we may be left with no universally endorsed rules of inference.[10] Certainly if logical nihilism is taken at face value as one of those competing logical theories, there would be no universally valid rules since it is a view that, by definition, denies that there are any good inference rules.

Notable about the characterization of both the approach and the problem in this discussion is the focus on inference *rules* that are not under dispute, rather than on arguments that are not under dispute. This obscures the fact that competing logical theories can agree that a given natural language argument is a good one while still disagreeing on (a) the correct logical form to ascribe to the sentences in the argument, (b) the validity of the inference pattern they ascribe to the argument, and/or (c) the correct account of validity.[11]

Both Aristotelian and classical quantified logics agree on the validity of the following well worn medieval example while disagreeing on the logical form of the sentences appearing in it:

Mortality

1. All men are mortal.
2. All Greeks are men.
3. All Greeks are mortal.

[9] See Payette and Wyatt (2018) for considerable elaboration on this point.

[10] See e.g. Russell (2018). Although it may not be the case that Russell's arguments apply to all logical laws (cf. Dicher, 2020), the ones that remain acceptable are very weak; maybe too weak to produce viable logical systems.

[11] This is a distinction that Martin and Hjortland acknowledge elsewhere, see Martin and Hjortland (2021).

According to classical FOL these sentences are universally quantified conditionals: $\forall x(Ax \rightarrow Bx)$, whereas the Aristotelian syllogistic categorizes them as universal affirmatives: AaB. Aristotelians explain the validity of **Mortality** by noting that it is of the valid argument structure *Barbara*, and further explain the validity of that structure with reference to the square of opposition. Classical FOL, in contrast, will show the validity of **Mortality** either by use of a deductive system or a model-theoretic semantics—in either case its validity will be explained in terms of more basic elements, whether basic deductive rules such as Universal Elimination and Universal Introduction, or definitions of satisfaction.

Advocates of classical FOL and independence-friendly (IF) FOL[12] will agree on the logical form and the validity of the inference in **Mortality**, while disagreeing on the correct account of validity—IF FOL usually appeals to game-theoretic definitions of consequence (Hintikka and Sandu, 1997).[13]

The point here might be usefully summed up this way. To the extent that agreement between logical theories is needed it is needed for particular arguments, albeit more interesting ones than our example of **Mortality**. The explanations of the acceptability of these arguments offered by competing logical theories are scientific hypotheses, which must be evaluated as part of the general process of theory choice.

Even logical nihilists, who maintain that there are no logical laws—no argument structures that are always and everywhere valid—can agree on the acceptability of particular arguments, though their explanations of the acceptability would have to be material and particular rather than purely structural. Of course there will be arguments the validity of which is under dispute, and nothing yet said here precludes the possibility that among those are some arguments that are essential to processes of theory choice. In the next section then we will consider more directly the nature of logical inquiry.

3 Dissolution

An alternate way to address—or more accurately dissolve—the background logic problem focuses, as gestured at above, on a more realistic model of theory choice in logic. In short, the view is this: the idea that the process of theory choice requires a background logical theory to proceed is a hangover from exceptionalism. The seeds

[12] The key feature of independence-friendly logics is a decoupling of the syntactic function of scope from the semantic feature of dependence, allowing that one quantifier could fall within the syntactic scope of another without the assignment of values to variables bound by the second quantifier being dependent on the assignment of values to variables bound by the first quantifier. The resultant logic has greater expressive power than standard FOL. See Tulenheimo (2022) for details and discussion of natural language applications.

[13] There is also a substantive theoretical disagreement between model theoretic and proof theoretic treatments of consequence for classical FOL itself, though this disagreement will not have any consequences for appraising arguments.

of this more realistic model can be found in our discussion above, but our goal here we lay it out more systematically.

First of all, let us consider what exactly must be taken for granted in order for the process of logical theorizing to get started. We do indeed need to take for granted that some arguments are good ones (namely the ones we are making as we proceed). But we need not take for granted a particular background logical theory to do so. This was the lesson of our discussion in Sect. 2.2.

To illustrate further, let us take a toy case in which we depend on a number of arguments which appear to be of modus tollens form. There are a number of logics which reject modus tollens as a valid logical law—paraconsistent logics, for example, or the logic offered in Yalcin (2012).

But—importantly—all those logics agree that some arguments that are of that form will be good ones. They may be understood as enthymemes to which other premises will need to be added to produce a valid argument, but as long as those additional premises are true in the situations in which we are using these arguments, our background reasoning will not be impacted. Alternatively these arguments may be non-deductively good: perhaps because of considerations from the study of reasoning more generally, or perhaps because of the subject matter—a materially good inference of the sort discussed in Norton (2003) and Brigandt (2010). In general logical theories need not disagree about the acceptability of specific arguments, even when they disagree as to the *explanation* of the goodness or badness of these arguments.

The idea that we are all using a logical theory, albeit perhaps an implicit one, seems to be a vestige of logical exceptionalism. If you think that human beings have some sort of logical intuition that allows them to perceive correct logical principles, then it makes sense to think of the acts of offering arguments and making inferences as applications of that logical intuition, and thus as invoking the logical theory which is out there awaiting discovery. And if you think that logical principles are part and parcel of linguistic competence, then of course offering arguments or making inferences reflects the background logical theory embedded in that competence. But for the anti-exceptionalist, the practice of offering and accepting arguments occurs prior to and independently of any theory that explains the goodness of arguments or justifies our acceptance and rejection of them, just as the practice of attempting to avoid dropping things made of glass—and indeed for that matter the successful creation of glass—precedes the creation and acceptance of physical theories explaining the relevant physical phenomena. We can do things—reason, create glass, predict that dropped things will break—in the absence of any theories explaining how we do them. Furthermore, we can often successfully do things even when under the influence of incorrect theories vis-a-vis the doing of those things—the practice of inoculating against disease precedes, by several centuries, the germ theory of disease transmission. To be sure, once these practices of reasoning (or glass-making, or inoculation) are adopted we begin to develop theories that explain their success, and these theories themselves may then influence our future practices, establishing as good new arguments (or glass-making or inoculation practices) on the basis of those theories. The point is just that we can get started without those theories.[14]

So it is not correct to say that we must presume a whole logical theory, or even particular schematic logical principles, in order to begin the process of theory-choice. Rather, we must assume that the specific reasoning we are using as we go on is reliable. And what we hope to get, among other things of course, out of the process of scientific inquiry is an explanation of why that reasoning is acceptable. Furthermore, the eventual explanation of that acceptability does not even require the acceptable argument to be valid in virtue of its form or necessarily truth preserving.

Secondly, the sorts of arguments offered in favour of adopting a particular logical theory are generally not the sorts of arguments that logical theories are theories of. One prominent anti-exceptionalist view, for example, is that logical methodology is characterized primarily by abductive reasoning or inference to the best explanation.[15] Now while there have certainly been some attempts to use formal tools to model abductive reasoning,[16] this kind of reasoning is not the subject matter of the various logical theories for which it is offered in support.

Another common view treats formal logic as a form of scientific modelling. There is no general consensus on the precise ways in which model-based science, in which one system is explored as a basis for understanding another system, is to be evaluated as successful or not, or even on how it generates knowledge. However there are some themes in the literature. Many models involve *abstractions* away from details that matter in the target system in order to create tractability—for example in ecological modelling the complexity of real systems is a barrier, and abstract ecological systems allow scientists to focus on specific aspects in isolation (Odenbaugh, 2005). Models that leave features of the real world out also allow scientists to provide *contrastive explanations*—for example modelling falling bodies in a vacuum can show the role of air resistance in the behaviour of falling bodies in the actual world (Hindriks, 2008). This same feature of models can also support *counter-factual reasoning* more generally (Bokulich, 2011). Finally it has been suggested that models provide explanatory scaffolding and thus a framework which can be applied to real systems or phenomena (Woody, 2008).[17] Choice between competing models is generally based on how useful they are for solving a particular problem or set of problems, and these sorts of judgments are not themselves the target of competing logical theories.

The usual framing of the background logic problem assumes that in order to have good arguments on which to base our logical theorizing we must have a background logical theory in place. The presumption is that the good arguments we begin with must be logically valid according to that background theory. But as we have seen there seems to be no good reason to think that this is true. Of course, once we start

[14] Thank you to an anonymous referee for pressing us on this point.

[15] For an overview of the view's popularity and an argument that it has limits as a methodology, see Hlobil (2021).

[16] For examples see Schurz (2008) and Thagard and Shelley (1997).

[17] We discuss the ways in which logical modelling might play these roles in Payette and Wyatt (2018).

the process of theorizing then we may well judge some of our initial arguments to be valid according to the theory we are developing, and that theory may lead us to identify further arguments as good because they're valid on that theory. But we can begin the process without any such theory in place.

Our approach to the problem combines elements of both the 'intra' and 'inter' framework replies, as characterized by Martin and Hjortland. One might worry, therefore, that even if we have successfully dissolved the putative problems faced by the inter-framework response the problems for intra-framework response remain. Recall that first, Martin and Hjortland objected that the intra-framework reply allows two research programs to use the same evidence to confirm their own starting logical theories. And second, they claimed that this approach does not preclude the possibility that a researcher would end up flip flopping between two logical theories L_1 and L_2 on the basis of the same data set—the idea being that L_1 together with the data gives one good reason to adopt L_2, but L_2 together with that same data gives one good reason to adopt L_1.

Our approach does not preclude the possibility that two research programs could both use the same evidence to confirm their own theory. But this should not be surprising to an anti-exceptionalist, since it is just a specific case of the general problem with the underdetermination of theory by evidence—something explicitly acknowledged by some looking at the methodology of logic as similar to that of natural science, cf. Hintikka (1968, 40). Other theoretical virtues can play an important role in theory-choice in these cases. The surprising result would be if logical theory choice was immune to such problems. Importantly, once one stops being an exceptionalist, there is no particular reason to think that what is assumed is the goodness of an argument *schema*, which can bring with it the suspicion that one is assuming parts of the theory without evidence.

Similar considerations should also give us some comfort with respect to the second problem for the intra-framework approach as well. If we begin the process not with a background logic but with a background set of arguments, then the possibility they are considering seems highly unlikely to occur, and if it does occur, then there are ways out of the loop.

If the evidence supports a logical theory L_1, then we have three components: the data set of pre-theoretic judgments of argument goodness and badness (Γ), the theory itself, and the reasoning (R_1) that connects Γ to L_1. This reasoning is not limited to deductive arguments; it might include abductive reasoning, inductive reasoning, model-based theorizing, etc. But let us assume, for the sake of our example, that R_1 includes at least some arguments which seem to be deductive—that is, the truth of the premises is taken to guarantee the truth of the conclusion.

The flip-flopping situation is one in which, having accepted L_1 on the basis of the initial data Γ together with this reasoning, we then, on the basis of L_1, have reason to accept L_2, where L_1 and L_2 are distinct and incompatible logical theories. This must mean that there is some body of reasoning R_2 that connects Γ to L_2. And again, this body of reasoning is not limited to, but may include, deductive arguments.

Furthermore for the flip-flopping to continue, there must be some body of reasoning, R_3, that connects Γ to L_1 on the assumption that L_2 is correct. R_3 might be the same as R_1, or it might not, since the process of adopting L_2 may lead us to reject some of our pre-theoretical judgments about what arguments are good ones.

Now there is nothing that can be said to rule this possibility out in principle. But as a practical matter, its worth asking how serious a concern this actually is. Consider for a moment those apparently deductive arguments that form part of R_2 and R_3. Neither L_1 or L_2 need to judge these arguments to be valid—i.e. good in virtue of their form. But presumably both theories would need to give some account of the goodness of these arguments, since otherwise it seems we would already have good methodological reasons to reject both theories—a logical theory which judges the very reasons we give for adopting it as bad sits on epistemologically shaky grounds at best. Furthermore, we have to assume, somewhat improbably, that the non-deductive reasons in R_2 and R_3 are themselves equally good and give us no grounds to choose.

Assuming that L_1 and L_2 agree on the goodness of any apparently deductive arguments in Γ, and agree on the goodness of any apparently deductive arguments in R_2 and R_3, their disagreements must lie in one of two places—either they disagree about the explanations of the goodness of those arguments, or they disagree on the goodness of some other arguments falling outside of these sets, or of course both. But in this case the situation is again simply one on which theory is underdetermined by evidence, and we will need to turn to other methodological principles in order to decide between L_1 and L_2. But as noted previously, this is exactly what the non-exceptionalist who takes theory choice in logic to be parallel to theory choice in the empirical sciences should expect to happen, at least in principle.

What does it mean to say that two logical theories can agree on the goodness of an specific argument without agreeing on whether it is valid? There will be cases where, to avoid paradox for example, we reject a pattern of argument previously thought to be valid. But in these cases it remains possible, indeed likely, that other arguments previously thought to be of that pattern are good ones: either because they are understood to be enthymemes with missing and true premises, or because they are understood to in fact have a different logical form than the problematic argument, or because they are understood to be materially good despite being formally invalid. Yalcin (2012) provides a suitable example here, since his proposed logic endorses so-called 'ordinary' cases of modus tollens while rejecting it as logically valid argument form.

Similarly, while paraconsistent logicians would generally deny the validity of disjunctive syllogism ($p \vee q, \neg q \vdash p$), they can also allow that this pattern of reasoning is generally reliable. Dialetheists, for example, can maintain that true contradictions are very rare.[18] As long as you have no reason (such as the appearance of a paradox) to suspect there is a true contradiction in the neighbourhood, the modus tollens pattern is materially reliable, despite its invalid form. And even a non-dialetheist paraconsistent logician could endorse a specific argument as good on non-deductive, material, grounds.

Even in cases where the disagreement is more substantial—say the disagreement between constructivist and classical mathematicians, we see a program to recapture as

[18] See Priest (1979, 231).

many of the classical results as possible. The underlying assumption of such a program must be that classical reasoning doesn't always steer you wrong as a practical matter—that a lot of the results are in fact correct—even if the means taken depends on invalid logical reasoning. Obviously in this case some arguments will in fact be given up entirely, and some results thus abandoned.

What about advocates of very small logics, like JC Beall? Beall's view concerns the scope of logicality—that is, he is arguing that only those arguments endorsed by subclassical logic are strictly speaking logically valid. But that is entirely compatible with the view that other arguments are materially good ones (Beall, 2018).

4 Conclusion

To recap then, our view is that the background logic problem does not arise once we recognize that the starting point for logical theorizing is not a provisional logical theory or even a collection of provisional logical principles, but instead a collection of arguments taken to be good. Proposed logical principles which unify parts of that collection, such as modus ponens, are hypotheses, not data, in the process of logical theorizing.

The arguments we make in support of a particular hypothesis are part of the collection of arguments we take to be good, but we can accept those arguments without *ipso facto* committing to a particular explanation of why they are good. Furthermore many of the reasons in favour of choosing one or the other logical theory will not be the subject matter of those theories in any case, but instead will be ones whose adequacy is judged on largely material grounds, such as abductive reasoning, or claims vis-a-vis the usefulness of a particular modelling practice.

References

Beall, J. (2018). The simple argument for subclassical logic. *Philosophical Issues, 28*(1), 30–54.

Bogen, J., & Woodward, J. (1992). Observation, theories and the evolution of the human spirit. *Philosophy of Science, 59*(4), 590–611.

Bokulich, A. (2011). How scientific models can explain. *Synthese, 180*(1), 33–45.

Bokulich, A. (2020). Towards a taxonomy of the model-ladenness of data. *Philosophy of Science, 87*(5), 793–806.

Boyd, N., & Bogen, J. (2025) Theory and observation in science. In E. N. Zalta & U. Nodelmann (Eds.), *The stanford encyclopedia of philosophy*. https://plato.stanford.edu/archives/spr2025/entries/science-theory-observation/

Brigandt, I. (2010). Scientific reasoning is material inference: Combining confirmation, discovery, and explanation. *International Studies in the Philosophy of Science, 24*(1), 31–43.

Cartwright, N. (1983). *How the laws of physics lie*. Oxford University Press.

Dicher, B. (2020). Requiem for logical nihilism, or: Logical nihilism annihilated. *Synthese, 198*(8), 7073–7096.

Evans, A. (1976). Causation and disease: The Henle-Koch postulates revisited. *The Yale Journal of Biology and Medicine, 49*(2), 175.
Feyerabend, P. (1981). Science without experience. In *Philosophical papers I: Realism, rationalism and scientific method* (pp. 132–136).
Finn, S. (2019). The adoption problem and anti-exceptionalism about logic. *The Australasian Journal of Logic, 16*(7), 231–249.
Haack, S. (1976). The Justification of Deduction. *Mind, 85*(337), 112–119.
Hindriks, F. (2008). False models as explanatory engines. *Philosophy of the Social Sciences, 38*(3), 334–360.
Hintikka, J. (1968). Epistemic logic and the methods of philosophical analysis. *Australasian Journal of Philosophy, 46*(1), 37–51.
Hintikka, J., & Sandu, G. (1997). Game-theoretical semantics. In J. van Benthem & A. Meulen (Eds.), *Handbook of logic and language* (pp. 361–410). North-Holland.
Hjortland, O. (2017). Anti-exceptionalism about logic. *Philosophical Studies, 174*(3), 631–658.
Hjortland, O. (2020). Disagreement about logic. *Inquiry: An Interdisciplinary Journal of Philosophy*, 1–23.
Hjortland, O., & Martin, B. (2023). Evidence in logic. In M. Lasonen-Aarnio & C. Littlejohn (Eds.), *Routledge handbook of the philosophy of evidence* (pp. 467–482). Routledge.
Hlobil, U. (2021). Limits of abductivism about logic. *Philosophy and Phenomenological Research, 103*(2), 320–340.
Kuhn, T. (1962). *The structure of scientific revolutions*. The University of Chicago Press.
Lombard, M., Pastoret, P.-P., & Moulin, A. (2007). A brief history of vaccines and vaccination. *Revue Scientifique et Technique-Office International des Epizooties, 26*(1), 29–48.
Maddy, P. (2002). A naturalistic look at logic. *Proceedings and Addresses of the American Philosophical Association, 76*(2), 61–90.
Martin, B. (2020). Identifying logical evidence. *Synthese, 198*(10), 9069–9095.
Martin, B., & Hjortland, O. (2021). Logical predictivism. *Journal of Philosophical Logic, 50*(2), 285–318.
Martin, B., & Hjortland, O. (2022). Anti-exceptionalism about logic as tradition rejection. *Synthese, 200*(2), 1–33.
Norton, J. (2003). A material theory of induction. *Philosophy of Science, 70*(4), 647–670.
Nutton, V. (1983). The seeds of disease: an explanation of contagion and infection from the Greeks to the renaissance. *Medical History, 27*(1), 1–34.
Odenbaugh, J. (2005). Idealized, inaccurate but successful: A pragmatic approach to evaluating models in theoretical ecology. *Biology and Philosophy, 20*, 231–255.
Payette, G., & Wyatt, N. (2018). Logical particularism. In N. Pedersen & J. Wyatt (Eds.), *Pluralisms: Truth and logic* (pp. 277–299).
Priest, G. (1979). The logic of paradox. *The Journal of Philosophical Logic, 8*(1), 219–241.
Priest, G. (2014). Revising logic. In P. Rush (Ed.), *The metaphysics of logic* (pp. 211–233). Cambridge University Press.
Reid, M. (2013). *Law and piety in medieval Islam*. Cambridge University Press.
Routley, R. (1980). The choice of logical foundations: Non-classical choices and the ultralogical choice. *Studia Logica, 39*(1), 77–98.
Russell, G. (2015). The justification of the basic laws of logic. *Journal of Philosophical Logic, 44*(6), 793–803.
Russell, G. (2018). Logical nihilism: Could there be no logic? *Philosophical Issues, 28*(1), 308–324.
Schurz, G. (2008). Patterns of abduction. *Synthese, 164*, 201–234.
Thagard, P., & Shelley, C. (1997). Abductive reasoning: Logic, visual thinking, and coherence. In *Logic and scientific methods: Volume one of the tenth international congress of logic, methodology and philosophy of science* (pp. 413–427). Springer.
Tulenheimo, T. (2022) Independence friendly logic. In E. N. Zalta & U. Nodelmann (Eds.), *The stanford encyclopedia of philosophy*. https://plato.stanford.edu/archives/fall2022/entries/logic-if/

Williamson, T. (2013). *Modal logic as metaphysics*. Oxford University Press.
Wimsatt, W. (2007). False models as means to truer theories. In *Re-engineering philosophy for limited beings: Piecewise approximations to reality*, Harvard University Press.
Woody, A. (2008). How is the ideal gas law explanatory? *Science and Education, 22*, 1563–1580.
Wright, C. (1986). Inventing logical necessity. In J. Butterfield (Ed.), *Language, mind and logic*. Cambridge University Press.
Yalcin, S. (2012). A counterexample to modus tollens. *Journal of Philosophical Logic, 41*(6), 1001–1024.

Logical Correctness for Anti-exceptionalism via Models

Jonas R. Becker Arenhart

Abstract Logical anti-exceptionalism has been frequently defined on negative terms: it is a view according to which logic is *not* an exceptional discipline. More positively, anti-exceptionalism sees logic as similar to other empirical sciences. That leaves a lot of space to characterize what the view actually (positively) amounts to. We propose that one fruitful way to do that is through the so-called 'logic as models' view. One of the benefits of doing so concerns the crucial notion of a true or right logic. By framing anti-exceptionalism in terms of models, we are in a better position to borrow from the philosophical literature on models to articulate views of how a model/theory relates to its target, accounting for what sense can be made of the very idea of a correct theory. We discuss two such accounts, and how anti-exceptionalism may be impacted by them.

1 The Background

Logical anti-exceptionalism is broadly understood as the view that logic is not a special discipline, with a *sui generis* methodology, epistemology, and metaphysics; rather, anti-exceptionalism proposes that logic is continuous with empirical sciences (Hjortland, 2017; Martin and Hjortland, 2022). Clearly, this is a general slogan, and it is not enough to unequivocally determine what the view amounts to in details. Unless one confers a more determinate aspect to this idea of continuity with science, the view will easily lead to incoherent features being attributed to it by different authors (see Rossberg and Shapiro (2021) for general critical comments). There is, however, a positive aspect to this flexibility: anti-exceptionalism will allow for some completely different strategies to be employed when it comes to make the approximation between logic and science, so that different and even incompatible versions of anti-exceptionalism may be explored, with pros and cons of each such approach open to evaluation.

J. R. Becker Arenhart (✉)
Federal University of Santa Catarina, Florianópolis, Brazil
e-mail: jonas.becker2@gmail.com

S. Molick (ed.), *Demarcating Logic and Science*, Studies in Applied Philosophy, Epistemology and Rational Ethics 80, https://doi.org/10.1007/978-3-032-25674-4_4

Despite all the flexibility allowed by its general characterization, however, logical anti-exceptionalism has been mostly formulated as an approach to *the epistemology of logic*; its stated aim is to overcome the limitations of so-called 'traditional *a priori* epistemologies' for logic,—the appeal to rational intuition or to analyticity—, which are considered limited or problematic in themselves and are said to be unable to provide for a method of logical theory choice (see in particular the discussion in Martin (2021)). In fact, according to this standard approach to anti-exceptionalism, the biggest epistemological challenge for logicians, it seems, is *to choose* a logical theory/theories,[1] among the many on offer, that actually present themselves as candidates for the correct theory accounting for validity *simpliciter* (Priest, 2016; Hjortland, 2017; Martin, 2021). Analyticity and rational intuition,—the traditional *a priori* methods—, just provide no resources for that task (see Read (2019) and Arenhart (2024) for a discussion).

One of the ways to present anti-exceptionalism consists in viewing the use of *abduction* as the major link between logic and empirical science (Priest, 2014; Priest, 2016; Hjortland, 2017; Martin, 2021). This is not the only way, but it still has major proponents. The idea, in a nutshell, is that both scientists and logicians look for the theory best instantiating the virtues any good candidate for a true theory should have, such as *adequacy to the data*, simplicity, explanatory power, generality in scope, predictive power, and so on.[2] In particular, *adequacy to the data* is understood as the most important feature, and by embracing the possibility of failing to account for the data, it is assumed that logical theories are open to revision when confronted with new evidence, just like empirical theories are (and this possibility of future revision is at least contemplated, even for those happy with the actual logic, whatever it is). In this case, the use of abduction has been praised as the much needed new epistemology for logic, one that meshes well with the very idea motivating anti-exceptionalism.

Of course, there are many difficulties with that view,—to put it mildly—, the explanation of what the data of logic are being one of the most pressing ones (see the criticism in Hlobil (2020), and Arenhart (2024)). Those difficulties have led some anti-exceptionalists to abandon the common use of abduction as the major link between logic and science (see Martin and Hjortland (2022), and footnote 2 of this chapter, concerning logical predictivism). What has *not been given up*, however, is the plan of providing some method for logical theory choice, with the basic underlying assumption that some one theory, at least, must be getting things right about logical validity. In other words, one must still look for the correct theory/theories that will be the one responsible to give us logical knowledge.

[1] The abductive method does not prejudge on the monism vs pluralism debate.

[2] *Logical predictivism* is one form of anti-exceptionalism based on the idea that one should choose the logical theory/theories best predicting valid inferences, where prediction is based on which inferences a theory validates (see Martin and Hjortland (2021) for an exposition). Although it is said to avoid the problems of abduction, it seems to be a form of abduction in disguise, given that this merely shifts the focus to one of the virtues present in the abductive process (predictive power), and still, cannot avoid the burden of having to tell when a prediction is well succeeded, i.e. when it does account for the facts being predicted.

In this chapter, we shall investigate a little deeper into the notion of the *correction* of a logical theory, according to some clues found in standard approaches to anti-exceptionalism. By being an attempt to improve the epistemic situation of logic by bringing it closer to empirical science, there is a sense in which one may benefit from the kind of discussion on correctness of a theory that has already been advanced in the philosophy of science literature itself. In other words, given that *logical* theories are similar to scientific theories, the account of the conditions for a theory to be correct should be similar too, and, if that is really the case, we may benefit from discussions already available in the philosophy of science literature.

Our strategy is the following one. Given that the standard accounts of theories in science take them to be characterized as collections of models (see Krause and Arenhart (2017), Frigg (2022), for models in science), or, even if that is not the case, that models mediate between theories and the data, it seems that the literature on scientific models may give us some interesting hints regarding such a relation. Furthermore, the idea that logical theories are models is not new; in fact, it has been developed already in the so-called 'logic as models' approach, and it has been waiting for anti-exceptionalist explorations until recently (see Cook (2010), Shapiro (2014)). So, in a sense, what we propose is that we could just advance a little more on the logic as models approach, with regard to the idea of logical correctness.

The relation of a model to its target is a very complex one, and it may be approached through distinct perspectives. We shall focus on the very problematic notion of logical data and the claim that theories should grant that such data are being taken into account as a central starting point to distinguish between two approaches to logical epistemology, mirroring two accounts found in the philosophical literature on models: the representationalist approach and the artefactual approach (see Knuuttila (2011), and Knuuttila and Boon (2011), for the terminology and further discussion, see Oliveira (2022) for a radical version of the view). Throughout this paper, we shall alternatively name them 'the Sherlock Holmes account', and 'the Einstein account', respectively (the terminology comes from de Ronde (2020), who advances it in a different context). In a nutshell, the accounts differ on the way they treat the data that a theory must account for. According to the Sherlock Holmes account, one must advance a theory to fit the data that is available beforehand, so that data comes first, and theories dispute to account for them. According to the Einstein approach, on the other hand, theorizing and conceptualizing the data are intertwined processes, and all encounters with the data are mediated by the theory. It will result that many important problems in logic will be addressed very differently depending on what stance one takes concerning these approaches to logical correction. It will become evident that most versions of anti-exceptionalism have been framed in terms of the Sherlock approach, with the Einstein approach remaining unexplored so far.

We shall develop the remainder of the paper as follows. In Sect. 2, we briefly remind the reader of what the logic as models approach amounts to. In Sect. 3, we discuss both the Sherlock approach and the Einstein approach in general lines. We indicate, by looking at the relevant literature, that most anti-exceptionalist approaches classify as the first kind of approach. In Sect. 4, we discuss some new perspectives that an Einstein-like approach may open to the epistemology of logic, and its close relation to the practice of logicians. We conclude in Sect. 5.

2 Logic as Models

Given that it is *logical theories* that may be correct, or fail to be correct, according to the anti-exceptionalist approach that we are investigating here, it would be very important for a fruitful discussion that we could start with a characterization of what a logical theory actually is. This strategy, again, is not new to logical discussions, but is taken from the philosophical discussion on science, which, as many people are quick to recognize, is completely theory-centered, in the sense that every aspect of the discussion concerning philosophical problems in science depends on an account of scientific theories. It is not different with logic, as we shall see, and clarity regarding the nature of theories is of utmost importance for a better appreciation of answers to most problems in the philosophy of logic. Here, we do not have space to discuss every detail of that matter, so we shall confine ourselves to the minimum necessary for our purposes.

What we find concerning this theme in the anti-exceptionalist literature is the recurring claim that logical theories are theories about logical properties, in particular, about validity (see Priest [2016, Sect. 2.4], Hjortland [2019, Sect. 2]). As Hjortland ([2019, p. 252]) puts it, a logical theory is not to be confused with a formal system, i.e., a "logical theory is not a proof system or a formal semantics with a consequence relation"; such formal systems do indeed contribute to our understanding of validity, but are not themselves a theory, according to this standard approach. A formalism, if it is to be a theory, "requires an application, and it is the logical theory that states what the application is" (Hjortland [2019, p. 253]). The application that makes a formalism into a *logical* theory, we are told, is application to validity in natural language, also known as 'validity *simpliciter*'. This characterization of a theory is a bit confusing, because it characterizes a logical theory as a formalism plus an application, while the application is explicitly defined by the theory, and so, requires the theory completely characterized beforehand. But what makes this characterization of a logical theory really opaque and not completely useful for some theoretical work is the fact that this only tells us what a logical theory is *about*, not exactly how it is articulated. Hjortland ([2019, p. 254]) provides additional important clarifications when he divides logical theories into metalinguistic and non-metalinguistic approaches; the former considering that validity is a linguistic property, while the latter considering that validity is a property of the non-linguistic world. That distinction is very useful, but it still deals only with what such theories are about, not with what they are.

Priest (2016) makes similar remarks, but now focuses on what are the goals of a logical theory, when he claims that such a theory provides "an account of the behaviour of certain notions (some of which are non-observational) and their interconnections" (Priest [2016, p. 39]). Luckily, we can improve on the specific topic of what a logical theory is by embracing the characterization of theories as provided by the logic as models approach.[3] The idea that logical theories may be understood as models, in a similar sense as the latter are understood in science, is not exactly new. The most well-known advocates of the view are Cook (2010) and Shapiro (2014). According to the latter:

> I propose that a formal language is a *mathematical model* of a natural language in roughly the same sense as, say, a collection of point masses is a model of a system of physical objects, and a Turing machine is a mathematical model of a person following an algorithm, or perhaps a computing device. In other words, a formal language displays certain features of natural languages, while ignoring, simplifying, or idealizing other features [...]. (Shapiro [2014, p. 46])

In terms of the general characterization of anti-exceptionalism, this looks like a promising start when it comes to an approximation of logic with science, in the sense that models characterizing theories in logic are not different in nature from models in science. The first nice advantage of using models, as the quote by Shapiro evidences, is that one may benefit from the fact that discussions on the nature and uses of models recognize right from the start the roles played by idealization, abstraction, and simplification, when it comes to theorizing and evaluating the merits of a theory. As it is widely recognized in discussions about models, the 'perfect model', in the sense of a model reproducing each and every detail of its target, is just not available.

The very fact that a logical theory, conceived as a model, idealizes and abstracts from natural language should already act as a kind of indication that the relation between a theory and its target will not be completely straightforward translation from natural language to logic; the view also highlights the fact that theorizing is a goal-oriented activity; what gets in the model and what stays out depends on specific purposes settled in the very act of theorizing, in the way we frame the problem to be dealt with. As Shapiro makes clear:

> With mathematical models, which features one focuses on, which are idealized, and which are ignored, depends on the purposes at hand, on why one is developing a model in the first place. Here, of course, our goal is to shed light on the relation (or relations) of logical consequence, and perhaps the norms for deductive reasoning and regulating beliefs to maintain consistency. So, presumably, in developing a logic-model, we should focus on and idealize those features of natural language that bear on deductive reasoning, or on regulating our beliefs for consistency, whatever those features may be. (Shapiro [2014, p. 47])

That suggests that there is a kind of selection of targets for specific purposes; the correctness of a model is not an all or nothing matter, with models involving a lot of distortions and simplifications of the target. We should contemplate the possibility that providing for a logical theory is a more convoluted activity than a simple description of features of transparent natural language. There are some accounts of scientific models that can help us approach logic by those directions.

---

[3] It is always good to remember that models in this sense are not to be confused with the models in the model theoretic approach. The latter are themselves part of the modelling activity that is involved in providing a model of validity, they act as models of the truth values and their relation to the connectives, for instance.

3 Models and the Targets

Once the notion of theory involved in the discussion is roughly understood as a model—and we are laboring under no illusion that we have explored all the details of the view—, we may investigate further into the kind of relation that obtains between the model and its target. That is crucial for the task that philosophers of logic on an anti-exceptionalist perspective see as facing them today, that is, to provide an account of logical theory choice. As mentioned before, theory choice is thought to be the biggest epistemological challenge for the philosophy of logic today. We have many systems of logic available, it is thought, and probably only one, or at most a few among them, are correct. The notion of correctness plays the most important role here. With the characterization of theories as models, we may investigate further into the notion of correctness, which is crucial for the debate. As a first stab at the issue, correctness requires that a theory somehow adequately describes the data or the evidence, its target.

As a result of this form of formulating the problem, our next task consists in providing a description of how theories relate to the evidence, so that we may more properly judge whether a theory is correct or not. In this section, we shall present two distinct ways to do that. The first one is to be found already in the logical literature, somehow in disguise. The second one is not yet completely explored in the logical literature, but is already to be found in the discussions of models in science.

3.1 The Sherlock Holmes Approach

The first (and rather obvious) way to see the relation between the theory and the data is the following one: there are some evidences or data available beforehand, and our theory must conform to them if it hopes to account for the data. In the case of logic, given that it is a theory about the validity of inferences in natural language, the data will be some inferences that will seem to us as obviously deserving to be called 'valid', whatever the theory about validity we are endorsing to begin with. Any theory competing for our attention will have to account for those inferences. As an instance, Richard Routley ([1980, Sect. 4]) advances the following list of facts to be taken into account by any theory expecting to be seriously considered:

Fact (1) Much of our discourse is intensional (while classical logic is extensional).
Fact (2) Much of philosophical discourse is about the non-existent.
Fact (3) There are inconsistent non-trivial theories and inconsistent non-trivial situations (while classical logic is explosive in the face of inconsistency).

So, there are some pre-theoretical facts about validity and the behavior of logical connectives, and a theory must be built in order to account for them. In this sense, we first collect the facts (at least some of them, which are considered irrevocable), and after that, we theorize and elaborate a theory based on them. This will be called the 'Sherlock Holmes approach', as we have already anticipated, because it follows very closely the recommendation of Sir Arthur Conan Doyle's character Sherlock Holmes, who, in 'A Scandal in Bohemia' famously stated that 'It is a capital mistake to theorize before one has data'. According to this account, data and theory are, in a sense, disconnected to begin with (again, the terminology comes from de Ronde (2020)).

That leads us directly to a discussion of the relation between a logical theory and what it is a theory of. Remember, it is said that logical theories are attempts to describe *logical validity*. The idea seems to be that there is something to be characterized, which would count as the very target phenomena of logic, and that Graham Priest has called *logica ens*:

> [...] it is crucial to distinguish between logic as a theory (logic docens, with its canonical application), and what it is a theory of (logica ens). In the same way we must clearly distinguish between dynamics as a theory (e.g., Newtonian dynamics) and dynamics as what this is a theory of (e.g., the dynamics of the Earth). (Priest [2014, p. 215])

We then face the situation where we may judge how distinct theories fare in the task of describing the data of logic. Priest suggests something similar to the Sherlock Holmes approach, arguing that a theory of negation cannot be arbitrary, "[t]he theoretical object has to fit the real object" (Priest [2006, p. 76]). He goes on:

> If it is to be applied, an account of negation must be considered not just as an abstract structure, but as a theory of something, just as a geometry is a theory of physical space. And this will put substantial constraints on what an acceptable account is. (Priest [2006, p. 77])

Again, the data available beforehand constrain the theory. This relation is profitably approached by the understanding provided by the relationship between models and their targets, the entities that the models are said to represent. This is basically what is involved in philosophical discussions of the correct logic, or correct logics, and what would it take for a logical theory to be correct. Blake-Turner and Russell, for instance, advance the following account of what would it take for a logical theory to be correct: it would have to be *accurate according to the facts*. Considering the question regarding the validity of the famous law of explosion as an example, they describe the problem as follows:

> What would it take to settle the question? Two things seem clear: (i) if the superstructure for some theory of classical logic is accurate, then explosion is valid. And (ii) if the superstructure for some theory of LP is accurate then explosion is not valid. And more generally, if we could determine that the superstructure of a theory for a logic V was accurate, and explosion was among V's E-sentences, that would tell us whether explosion was valid tout court. (Blake-Turner and Russell [2021, p. S4863])

'Accuracy' here means basically an adequate description in the theory of logical features of natural language. If the facts about validity in natural language do involve only two truth values, a classical-looking behavior of the logical connectives, and validity of the law of explosion, then, classical logic, which advances precisely such a kind of description, is the correct one. If, on the other hand, natural language involves true contradictions, i.e. some sentences being true and false, and the connectives behave just as prescribed by **LP** (Graham Priest's *Logic of Paradox*), with failure of explosion, then, of course, **LP** is the accurate description (see Priest (2006) for more on **LP** and true contradictions). The idea is that there should be a perfect match between the theory and the 'facts' about natural language concerning notions such as truth, falsity, the meaning of the connectives and of logical consequence (see Blake-Turner and Russell (2021) for further discussion).

This view of the relation between a model and its targeted system is already quite widely explored in the philosophy of science literature, and it is related to investigations concerning how models represent (Frigg [2022, Chap. 6]). The traditional approach to logical theorizing as an attempt to describe the data of logic could benefit from accounting for logical theories in terms of models, given that this would bring in an epistemology that is already well explored in the philosophy of science literature. As Knuuttila and Boon summarize:

> According to the received wisdom, models give us knowledge because they represent their supposed external target systems more or less accurately, in relevant respects and to a sufficient degree. This kind of formulation already suggests that there is a special sort of relationship between a model and its target, which is most commonly analysed in terms of similarity or isomorphism. (Knuuttila and Boon [2011, p. 311])

The idea that logical knowledge is obtained once a theory accounts for the data, in the specific sense that the model (that stands for the theory here) is *similar* or *isomorphic* to the data suits quite well the abductivist approach (and the logical predictivist approach too), whose major claim is that knowledge is obtained by the selection of theories accounting for the data, and instantiating the distinguished set of pragmatic virtues to a high degree. Here, *similarity* between model and its target, or *isomorphism* between model and target, are two standard ways of accounting for the relation between models and targets in the philosophy of science literature. Roughly, the idea is that a model correctly captures its target when it is *similar*, in a sense, to the target, or, when it is *isomorphic* to a data structure, a structure representing the data of the target (see Frigg [2022, Chap. 6] for further details). This seems to square nicely with what abductivists have in mind when accounting for the relation of theory and evidence:

> According to this account of logical epistemology, logical propositions are not directly justified by intuitions or definitions, but rather logical theories are justified by their ability to best accommodate relevant data. In other words, logical theories are justified by abductive means. (Martin [2021, p. 9071])

Of course, we are not claiming here that this is how anti-exceptionalists have explicitly put the terms, but only that the literature in philosophy of science is already quite developed in such aspects. Exploring the already available accounts of this relationship between theory and data in the philosophy of science literature, such as isomorphism and similarity, may prove useful for those defending that logical theories represent more or less correctly the data (again, see also Frigg [2022, Chap. 6] for further details on scientific representation). The emphasis should be on the idea that logical knowledge is obtained once such a relation between data and theory is established. This puts the matter of logical knowledge in clearer terms, and clearly brings logic closer to a particular account of the workings of science.

3.2 The Einstein Approach

A usually discussed theme in the philosophy of science when it comes to the relation between data and theory is the theory ladeness of the data. To put it very briefly, the idea is that one cannot have data in a neutral fashion, every piece of evidence is already laden with some theoretical presupposition. This will give rise to the Einstein approach; the name comes from a very famous situation experienced in an encounter by the then young Werner Heisenberg and the more experienced Albert Einstein, where Einstein counseled to Heisenberg that 'it is only the theory which decides what you observe' (after that, Heisenberg discovered the famous uncertainty principles that go under his name in quantum mechanics).

The idea of theory ladeness may be nicely illustrated when we discuss it in the context of modeling in the sciences (so that we may also benefit from the literature on scientific models). This is important, given that modeling is seen as a central activity in scientific practice. This will also allow us to connect logical anti-exceptionalism with the logic as models approach in a direction different than the one explored previously.

One of the alternatives to the representational view includes the *models as epistemic tools approach*, or the *artefactual approach to models*, as proposed by Knuuttila and Boon (2011) and Knuuttila (2011). According to the view, in a nutshell, models are less like independently given entities, and more like artifacts: "scientific models could be usefully approached through their artefactual dimension as constructed entities that give theoretical interpretations of some target phenomena in view of particular epistemic purposes" (Knuuttila and Boon [2011, p. 310]). The view has a major impact on the epistemic dimension of science, given that it does not identify knowledge exclusively with the obtaining of a relation between two independent entities, the model and the target, as per the representationalist approach we have discussed previously. According to the artefactual approach, modeling incorporates epistemic dimensions in many distinctive stages, from *formulating* the appropriate questions and settling the goals of inquiry, *interpreting* the target in terms of scientific concepts available beforehand, and *constructing* the model in a reciprocating process between model and target, where the target is used to construct the model, and the model is used to shape the target (again, for the details, see Knuuttila and Boon (2011)).

In particular, what is more relevant for us here are two distinct features of the account of models as epistemic tools. First of all, the justification of such models is quite distinct from the representationalist view. The suggestion is that models have a kind of built-in justification, it incorporates parts of what has to be accounted for by the model already in the process of model building:

> It is important to note that models often carry with them initial justification that is due to the various ingredients they comprise, such as various renderings of empirical data and knowledge, theoretical assumptions, mathematical formalisms, and established modelling methods. (Knuuttila [2011, p. 269])

The idea of built-in justification comes from Marcel Boumans, who explains the idea in the following terms:

> I have argued that the model-building process is the integration of several ingredients in such a way that the result—the model—meets certain *a priori* criteria of quality. And because some of the ingredients are the generic data the model is supposed to explain, justification is built-in. (Boumans [1999, p. 91])

Before moving on, let us comment on the '*a priori* criteria of quality' mentioned by Boumans. This is not meant to indicate that some of the knowledge of data is afforded without need of experience or experiments—as per typical characterizations of the *a priori*—, but rather that the model is constructed already incorporating some of the data it is expected to explain, so that, in the end, success in explaining such data is granted right from the start; i.e. it is not surprising that the model accounts for some of the data, once they are ingredients in the construction of the model.

It is not difficult to see how these apply to the discussion of logic concerning the rule of explosion, to use a most prominent example once again. In the case of classical logic, the models already incorporate the fact that contradictions are not tolerable; typically, in the usual approach, each sentence receives only one of two mutually incompatible and jointly exhaustible truth values. The background assumption is that contradictions are always false, and the model of logical consequence, along with the theoretical concepts involved, is developed to reflect that. Classical models, however, may be modified in order to generate **LP** models. In such models (or, at least on some account of the construction of a version of **LP**; see Omori and Arenhart (2023) for further discussions), truth and falsity are not exclusive, with some sentences bearing both truth values. That is, the ingredients used to construct the models involve *truth value gluts*: this is not something we find out by using the **LP** models. In other words, **LP** succeeds in accounting for true contradictions because they are built-in (some versions of) the construction of the theory. The use of such ingredients to block explosion should not come as a surprise: the models were built precisely for that (in this sense, they instantiate a partial explanation for the success of the application of mathematics). Something similar and clear occurs in the case of Lukasiewicz three-valued logic. The idea that some sentences should be indeterminate, that they represent future contingent sentences, is used in the very construction of the model. It is not, as a representationalist account would require, that we first have a mathematical model, and only in a later stage of the investigation attempt to connect it to data already existing. No, the selection of the relevant data and the production of the model is completely intertwined process, according to the models as epistemic tools approach, so that the justification is at least partially built-in.

This leads us to the second important feature of the models as theoretical tools approach. Recall that not only part of the relevant data is used in the development of the model, but also that models contribute to the very theoretical interpretation of the phenomena to be dealt with, given some purposes (again, see Knuuttila and Boon [2011, p. 310]). The plan is that one does not address the phenomena without a background of assumptions and concepts. Rather, one interprets the phenomena in terms of the concepts provided by a model. This will lead to one evaluating very differently the idea of the data in logic. Let us go back, again, to the issue of whether the law of explosion is valid or not (the example discussed also by Blake-Turner and Russell (2021)). The question depends very much on whether one is open to the possibility of counter examples to it, which in most cases requires that some contradictions are seen as acceptable (or that one can, at least, formally accommodate them as tolerable in the model). But once one adopts a classical approach, the theoretical apparatus involved will simply not allow for such counterexamples. That makes good sense of the attitude of classical logicians towards inconsistent theories: given the background theoretical assumptions they use to interpret such data, contradictions cannot be true. On the other hand, by using a paraconsistent model of a sort, one is able to see such candidates for true contradictions as completely legitimate, and account for them as gluts in some cases, or as not prejudicial to the coherency of the system in general, at least. No wonder, then, that those using different frameworks will attribute different importance to such 'data': they see it through distinct theoretical apparatuses, and the impact in each case is different (see the detailed discussion in Arenhart (2022a)).

Clearly, this brief presentation is not meant to be a completely developed defense of the use of the artefactual account of models in logic. Rather, the plan is that it opens the possibility to completely reframe the debate concerning the data of logic, so that we avoid some of the pitfalls that a completely representationalist approach would have to face. Data is not neutral, it is theoretically elaborated modulo some theoretical background, and theories are not correct by being accurate to pre-theoretically given data, because part of the ingredients of a theory already involve using the very data they are supposed to model, and the understanding of what counts as data is also informed by the theory in a kind of bootstraping process. This will have a clear impact on the idea that there is one true logic: now, a logical theory contributes to the very data it is expected to model, and the very idea of a theory being true of something (the target) given independently of the theory seems inappropriate.

Overall, the artefactual approach, when brought to logic, has deep impact on the metaphysical underpinnings of the anti-exceptionalist view too. Notions of data, adequacy to the data, correctness, truth and validity simpliciter will have to be reassessed.

4 New Dimensions for Anti-exceptionalism

We have started this chapter by discussing how anti-exceptionalists are actually worried with the improvement of the epistemology of logic. Given the alleged inability of traditional methods of epistemology to deal with the actual situation in logic, new perspectives must be sought, and such epistemology will have to account for the actual diversity of logical theories. Research on this front, however, has focused on theory choice, with most of the literature emphasizing that a correct theory must be available once and for all. This is encapsulated in the Sherlock Holmes approach we presented earlier. Besides the difficulties that the approach offer, we think that focusing on theory choice and on the search for a true or correct logic would miss the opportunity for a real change in priorities in the philosophy of logic. In fact, by focusing on the search for the true logic/logics in the context of anti-exceptionalism, we are just pursuing the same traditional ends, but with different means. The change in the epistemology for logic should go deeper, we suggest, and the purely representationalist account does not deliver that additional step.[4]

What the models as epistemic tools suggest is that there is much more to scientific knowledge than just finding a theory that matches the data (and the approach indicates that doing so is ill guided). This lesson could be learned by philosophers of logic, so that one could recognize that we actually do have more logical knowledge than we recognize when we adhere to use of the abductive method with the Sherlock Holmes approach as the appropriate epistemology for logic (which, it seems, will leave us with no knowledge at all, as far as we do not seem to agree on the correct logic). Of course, this does not mean that searching for the one true logic (if one is a logical monist) or for the true logics (if one is a logical pluralist) is a completely idle activity. Doing so is part of what some members of the logician's community actually do, that part of the community that works in philosophy departments and that are on the 'search for the correct theory' mode; but focusing on that activity, and restricting knowledge attribution to the resolution of that problem blinds us to all other important activities that are actually successfully pursued under the rubric of logic. By shifting our attention to science, anti-exceptionalism has the chance to learn to recognize other dimensions of the research in logic that also deserve to be considered as contributing to the epistemic endeavor of science.

This is where the account of logical theories plays a most prominent role for the selection of the important problems for the philosophy of logic, and this is where the connection with science may play an illuminating role. The fact that models and modeling are a constitutive part of science has brought many fruits for the philosophical understanding of empirical science. Not the least one is the fact that by studying the workings of models in scientific practice, one gains a more realistic understanding of science itself, which is what philosophers of science are looking for, anyway. That kind of approach can also be profitably pursued by logicians willing to understand the nature of logic nowadays. This becomes clearer when we explore other dimensions of modeling in science and consider their impact in our understanding of logic. That is, it is a call for a somewhat more practice-oriented philosophy of logic.

Ben Martin, for one, has suggested that the epistemology of logic may benefit from contemplating the practice of logic, just as the epistemology of science and

---

[4] See also Tajer (2022) for another approach to anti-exceptionalism that also seeks to leave the idea of choice of a correct theory behind. Tajer's approach, however, is less concerned with the nature of models employed, and more related with the multiplicity of practices of logicians.

mathematics benefits from considerations deriving from practice; accounting for the practice is a feature of a desirable epistemology of a discipline:

> Any theory that fails to take seriously a significant portion of practice must be considered deficient. Inversely, any theory that accommodates important elements of practice has an advantage over competitors. (Martin [2021, p. 9077])

However, as we mentioned, this practice-oriented account has not, so far, been considered also in connection with the logic as model approach, even though they seem to be made one for another, and seems to be missing the richness of the different dimensions of the logical practice that transcend the disputes on the correct logic (see also Martin (2022)). We shall illustrate one aspect of how this practice goes way beyond the search for the one true logic, and may be accounted for in the lines of the more general approach to modeling offered by the artefactual account of models.

Recall the classical discussion in science by Knuuttila and Toon that merely representing is not enough to account for most of what is considered as epistemically valuable in science. Models serve a varied set of purposes that go much beyond that, and that philosophers' single-minded pursuit of correctness has largely ignored. As Knuuttila argued:

> In purely philosophical terms, the idea of grounding the knowledge we gain from models in accurate representation may seem impeccable, but then what other evidence could there be for the correctness of a model than what has been built into it, the results it gives, and the inferential links to other bodies of knowledge it establishes? To require, then, that we have knowledge first when some definite representational relationship between the model and some real target system has been forged leaves much scientific work unrecognized and its epistemic value unexplained. (Knuuttila [2011, p. 270])

That holds also for the representational Sherlock-style view of logic. The account in question does imply that the only way to obtain logical knowledge obtains when a model is isomorphic or similar to its target. Everything else that logicians do—and they do a lot of different stuff—, will not count as knowledge. The result, given that we do not have any final claim to the correct theory, is that we seem to know nothing about logic (e.g. logicians do not know techniques to obtain specific results as completeness and incompleteness theorems, they ignore how certain theories relate to each other, they do not know how to provide for proof systems closely related to a given model theory, and so on...). But that would be disastrous. That only shows that such new approaches to the epistemology of logic have failed on their task of accounting for the successful practice of logicians.

As Knuuttila and Boon indicate in the quote above, there are important epistemic dimensions that are present in the study of the relations between different bodies of knowledge. That passes unacknowledged when knowledge is just taken to mean correspondence with the facts. This is exactly what Curtis Franks claims concerning the goal of logical investigation:

> One should warm up to the trend of identifying logic with the specialized scientific study of the relationships among various systems and their properties. This is, after all, how logicians use the word. Our preference to ignore questions about a logic's correctness stems not only from an interest in exploring the properties of possible logical systems in full generality but

> also from an appreciation, fostered by the study of logic, that no one such system can have all the properties that might be useful and interesting. (Franks [2015, p. 163])

A philosophy of logical practice that really encompasses every dimension of logical knowledge should not be blind to the immense amount of logical knowledge that is available, and that would be inaccessible for us in case the only source of knowledge should be something along the lines of the Sherlock account. In this sense, anti-exceptionalism should give a more radical step ahead, and really let the practice orient it, guided by suggestions that there are important dimensions of knowledge that are left unaccounted for by such simple approaches as the abductive one.

In order to illustrate one further advantage of the approach being proposed here, we shall briefly consider part of the activity of logicians that has been ignored by the current anti-exceptionalist accounts, and that clearly involve epistemic dimensions in the sense described by Knuuttila and Boon, and Franks, in the passages we have quoted a few paragraphs before. We shall consider one aspect of modeling that is very important in science in general, and in logic in particular: the use of models for the purpose of mathematical generalization. The fact that the models used in providing for a logical theory are mathematical models opens up the gates for the most wild kinds of generalization. Doing so is part of the activity of logicians as well as mathematicians, and that can be done with no direct concern for the search of the correct system of logic, while still involving clearly important epistemic dimensions.

Models are already seen as tools for generalization in general philosophy of science. Here is an account of how the fact that a model is formulated in mathematical terms can be used in favor of generalization:

> Mathematics, in this regard, is not merely an auxiliary tool for applying fundamental theories to specific empirical situations or phenomena. Rather, as Mary Hesse puts it, 'any particular piece of mathematics has its own ways of suggesting modification and generalisation' which, in turn, may take the place of the more traditional 'pointers towards further progress' [...] While mathematics thus affords considerable potential for exploration, whether or not its exploratory use in modeling will lead to overall scientific progress also depends on the presence, or absence, 'of a scientific community whose members are skilled in applying and modifying models and theories, and which, collectively, is able to arrive at determinations regarding the fruitfulness (or "progress") of new theoretical proposals' [...]. (Gelfert [2016, p. 83])

Clearly, what holds for the fundamental empirical theories also holds for logic, whose history seems to be one of increasing generalization ever since the first mathematical approaches have been available. As the quote indicates, the very fact that a theory has a mathematical formulation suggests by itself generalizations that may be explored by skilled researchers and eventually applied fruitfully again. For examples of the kind of generalization allowed by the mathematical framing of logical models, and which are developed sometimes with applications in mind, sometimes merely for the sake of mathematics, one need not go very far: consider different families of logic available in the literature, paraconsistent logics (such as da Costa's systems and the more general LFIs), the many-valued logics, whose history is a straightforward case of generalization from two-valued logic, constructive logics, modal logics (with all kinds of reading for the modalities), and so on. There is a successful story of generalization behind the development of such systems, and it would be simply inappropriate for an epistemology of logic not to count those facts as important epistemic achievements of logicians.

The impact of such considerations on the Sherlock account of theories and logical knowledge is considerable. When one recognizes that there are many important aspects of the activities of logicians that are simply not considered as involving epistemic dimensions when resorting to that view, it starts to lose its credentials, and the Einstein approach appears as more promising. Clearly, it is possible to seek for deeper positive impact in the epistemology of logic if we leave that final obstacle related to the dispute regarding the one true logic/logics behind.

5 Conclusion

As we have discussed in this chapter, although anti-exceptionalism is an attempt to bring a revolutionary twist to the epistemology of logic, its accounts of logical theory are very focused on the major claim that logic is busy describing a notion of logical consequence in natural language or in the world, a notion of *validity simpliciter* or *validity tout court* (as we have seen the discussion in Hjortland (2019), Blake-Turner and Russell (2021), and Priest (2014) with the notion of *logica ens*). The epistemology advanced by anti-exceptionalists in order to achieve the goal of properly describing how we achieve logical knowledge is connected with logical theory choice; we choose the system that best accounts for the data, and that theory will give us knowledge. This account assumes a clear separation between theory and data, and knowledge occurs when a matching relation between them obtains.

As we have discussed, that view is one possible way to go, but it seems to fail in taking into account the multiple dimensions of knowledge in science in general, and in logic in particular. So, we have suggested that advancing logical theories may be seen as the result of a more complex activity of modeling, similar to some accounts of modeling in science, which recognizes important dimensions that include more than just the matching relation of the model and the target. Rather, there is a complex relation between target and model where model and target inform each other, and the very understanding of what counts as data is theory-related. We have suggested that this account (the models as epistemic tools account) has the resources to address the activity of logicians investigating the mathematical properties of systems that are not exclusively related in the dispute for the correct logic. That opens the horizon for a more practice-engaged form of anti-exceptionalism. As a quick illustration, we have suggested that discussions concerning mathematical generalizations are frequent in logic, and have no place in the usual anti-exceptionalist account of logic. By bringing those dimensions to light, we hope, a more faithful account of the epistemology of logic will be achieved, and the difficulties that beset the actual approaches will be dissolved.

References

Arenhart, J. R. B. (2024). Abductivism as a new epistemology for logic? *Erkenntnis, 89*, 2397–2415.
Arenhart, J. R. B. (2022). Understanding logical evidence, with lessons from the paradoxes. *Journal of Philosophical Research, 47*, 177–194.
Blake-Turner, C., & Russell, G. (2021). Logical pluralism without the normativity. *Synthese, 198*(Suppl. 20), S4859–S4877.
Boumans, M. (1999). Built-in Justification. In: M. S. Morgan & M. Morrison (Eds.), *Models as mediators. Perspectives on natural and social science* (pp. 66–96). Cambridge University Press.
Cook, R. (2010). Let a thousand flowers bloom: A tour of logical pluralism. *Philosophy Compass, 5*(6), 492–504.
de Ronde, C. (2020). Quantum theory needs no 'interpretation' but 'theoretical formal-conceptual unity' (or: escaping Adán Cabello's "Map of Madness" with the help of David Deutsch's explanations). Phil. Sci Archive http://philsci-archive.pitt.edu/17688/.
Franks, C. (2015). Logical nihilism. In A. Hirvonen, J. Kontinen, R. Kossak, & A. Villaveces (Eds.), *Logic without borders* (pp. 147–166). Walter de Gruyter.
Frigg, R. (2022). *Models and theories: A philosophical inquiry*. Routledge.
Gelfert, A. (2016). *How to do science with models. A philosophical primer*. Springer.
Hjortland, O. T. (2017). Anti-exceptionalism about logic. *Philosophical Studies, 174*, 631–658.
Hjortland, O. T. (2019). What counts as evidence for a logical theory? *Australasian Journal of Logic, 16*(7), 250–282.
Hlobil, U. (2020). Limits of abductivism about logic. *Philosophy and Phenomenological Research*. https://doi.org/10.1111/phpr.12707
Knuuttila, T. (2011). Modelling and representing: An artefactual approach to model-based representation. *Studies in History and Philosophy of Science, 42*, 262–271.
Knuuttila, T., & Boon, M. (2011). How do models give us knowledge? The case of Carnot's ideal heat engine. *European Journal for Philosophy of Science, 1*, 309–334.
Krause, D., & Arenhart, J. R. B. (2017). *The logical foundations of scientific theories: Languages, structures, and models*. Routledge.
Martin, B. (2021). Identifying logical evidence. *Synthese, 198*(10), 9069–9095.
Martin, B. (2022). The philosophy of logical practice. *Metaphilosophy, 53*, 267–283.
Martin, B., & Hjortland, O. T. (2021). Logical predictivism. *Journal of Philosophical Logic, 50*, 285–318.
Martin, B., & Hjortland, O. T. (2022). Anti-exceptionalism about logic as tradition rejection. *Synthese, 200*(148), 1–33.
Oliveira, G. S. (2022). Radical artifactualism. *European Journal for the Philosophy of Science, 12*(36). https://doi.org/10.1007/s13194-022-00462-0
Omori, H., & Arenhart, J. R. B. (2023). Change of logic without change of meaning. *Theoria, 89*(4), 414–431.
Priest, G. (2006). *Doubt truth to be a liar*. Oxford University Press.
Priest, G. (2014). Revising logic. In P. Rush (Ed.), *The metaphysics of logic* (pp. 211–223). Cambridge University Press.
Priest, G. (2016). Logical disputes and the a priori. *Princípios, 23*(40), 29–57.
Read, S. (2019). Anti-exceptionalism about logic. *Australasian Journal of Logic, 16*(7), 298–318.
Rossberg, M., & Shapiro, S. (2021). Logic and science: Science and logic. *Synthese, 199*, 6429–6454.
Routley, R. (1980). The choice of logical foundations: Non-classical choices and the ultralogical choice. *Studia Logica, 39*(1), 77–98.
Shapiro, S. (2014). *Varieties of logic*. Oxford University Press.
Tajer, D. (2022). Anti-exceptionalism and methodological pluralism in logic. *Synthese*, *200*, Article 195.

Disentangling Scientific Realism from Anti-Exceptionalism About Logic

Jonas R. Becker Arenhart, Raoni Arroyo, and Ederson Safra Melo

Abstract Scientific realism is, currently, one of the most well-entrenched background assumptions of some relevant versions of anti-exceptionalism about logic. We argue that this is a sort of sociological contingency rather than a metaphilosophical necessity. Drawing parallels with the metaphysics of science (as applied to quantum foundations), we try to bring the realist assumptions of anti-exceptionalism to light, to undermine the necessary connection between realism and anti-exceptionalism, briefly exploring the possibility of adopting antirealism as the background default view of science instead.

1 Introduction

It is an undisputed fact that there is an infinity of distinct formal systems of logic. They are all presented through the use of various mathematical formulations, either in terms of a semantic approach, or else in proof-theoretic versions, and, in a sense,

J. R. Becker Arenhart
Department of Philosophy, Universidade Federal de Santa Catarina (UFSC), Florianópolis, Brazil

Graduate Program in Philosophy, Universidade Federal do Maranhão (UFMA), São Luís, Brazil

J. R. Becker Arenhart · R. Arroyo
Research Group on Logic and Foundations of Science (CNPq), Florianópolis, Brazil

J. R. Becker Arenhart · R. Arroyo · E. S. Melo
Research Group on Philosophy of Logic and Language (GFILL/CNPq), São Luís, Brazil

R. Arroyo
Graduate Program in Philosophy, Universidade Federal de Santa Catarina (UFSC), Florianópolis, Brazil

Centre for Logic, Epistemology and the History of Science, Universidade Estadual de Campinas (UNICAMP), Campinas, Brazil

E. S. Melo (✉)
Department of Philosophy, Universidade Federal do Maranhão (UFMA), São Luís, Brazil
e-mail: saframelo@gmail.com

S. Molick (ed.), *Demarcating Logic and Science*, Studies in Applied Philosophy, Epistemology and Rational Ethics 80, https://doi.org/10.1007/978-3-032-25674-4_5

these systems differ in the answer they provide for which inferences are considered to be licit. Facing a problem where logical tools are required, we find ourselves having to choose an appropriate logical theory. One such context, according to philosophers of logic, is quite special, it involves addressing the substantial problem of what *actually* follows from what, the so-called *canonical application of logic*; in this case, we are in the business of having to choose the *correct* account of logical consequence for inferences drawn in natural language. That is, we have to spot which among the many systems available correctly describes what follows from what in an extra-systematic sense (see Haack, 1978, Chap. 2). So, this view assumes that the correct account of logical consequence corresponds to some informal, or extra-systematic, conception of logical consequence. As Martin (2021) puts it:

> Probably the most prominent and philosophically interesting example of such theory choice is over logics that aim to not just solve a certain technical or philosophical problem, but aim to provide a general account of validity—*what propositions follow from what*. Multiple logics have been proposed as viable candidates to capture validity: paraconsistent logics, intuitionistic logics, paracomplete logics, quantum logics, and of course classical logic. Given that these various logics license different rules of implication, often with significant repercussions, it clearly matters which logic we ultimately endorse. (Martin, 2021, p. 9070).

One frequently finds the claim in current philosophy of logic literature that, facing this specific version of the problem of theory choice, the traditional accounts of the epistemology of logic just do not seem to be up to the task. Traditional epistemology of logic is said to rely on a priori methods—rational intuition or semantics of logical vocabulary—that, besides being problematic in themselves, just cannot account for the fact that logic is now thought to be revisable, i.e. that we may change our view as to which logical theory correctly describes validity depending on the evidence we have at our disposal at a given time (see Priest, 2016; Hjortland, 2017; Martin, 2021). As a result, logic, taken as a discipline, lacks an appropriate epistemology.

But not all is lost. There is a new tendency in philosophy of logic, going under the name of *anti-exceptionalism about logic* (AEL, hereafter), that tries to fill this gap by bringing logical knowledge closer to the knowledge of empirical sciences. AEL is the view that logic is not 'exceptional' (as the name claims) with respect to other sciences; that means, among other things, that there is a kind of 'common epistemology' for logic and for empirical sciences. According to the oft quoted characterization of AEL:

> Logic isn't special. Its theories are continuous with science; its method continuous with scientific method. Logic isn't a priori, nor are its truths analytic truths. Logical theories are revisable, and if they are revised, they are revised on the same grounds as scientific theories. These are the tenets of anti-exceptionalism about logical theories [...]. (Hjortland, 2017, p. 632)

It is not an accident that in this epistemological proposal, the focus has been put on logical theory choice. Particularly, the idea is to bring the methodology of scientific theory choice to logic as an attempt to boost the latter's epistemology. The plan is quite simple: science has an incredible track record of success; let logic benefit from that as well! By using a method of theory choice similar to the one employed in science, we not only recognize that logical theories are revisable,[1] but also employ some theory choice methods whose reliability in delivering the goods has an incredible success rate (see Priest, 2014, 2016; Hjortland, 2017; Martin, 2021).

In this paper, we explore one specific aspect of AEL's proposal of addressing the idea of continuity between logic and science in terms of the continuity in methods of theory choice or theory appraisal in general. Our major claim is that AEL—as characterized in most of current literature—takes for granted that there is a unified and completely consensual approach to what are the methods and aims of science. This supposition results in AEL subscribing to a straightforward realist description of the workings of science.[2] Given this view, it would be just a matter of transferring the use of those methods to logic, and at least part of the seeming stalemate of theory choice in the philosophy of logic would be solved. But one can never forget that realism is not the only way to frame the aims and proper understanding of science.

As we shall suggest, continuity with science can be approached through different and incompatible lines, with the proper aims and methods of science being a highly disputed topic in the philosophy of science literature. As a result, the idea of continuity between logic and empirical science is not a simple matter, and framing it as a question of theory choice by use of abduction or some focus on the explanatory power of theories is just one possibility among many. It is not to be expected that all of those different ways to frame the similarity between science and logic would be congenial to the aims of actual major formulations of AEL. That, by itself, is not an argument against AEL, but against taking for granted that there is only one way to link logic with other branches of science. As even some anti-exceptionalists are now recognizing, the characterization of AEL in terms of abduction and logical theory choice faces difficult challenges; our hope is that by bringing some of the underlying commitments of the view to light, more appropriate approaches may be sought.

The structure of this chapter is as follows. In Sect. 2, we address the question of which view of science AEL seems to be taking for granted. In Sect. 3, we borrow some instructive illustrations from the case of quantum mechanics and the metaphysics of science, where metaphysics tries to obtain some epistemic warrant from the success of science. The case is instructive because it reflects a common pattern in attempts of naturalization or approximations with science in general. In Sect. 4, we bring to light some general lessons that AEL may learn from the previous discussions. We conclude in Sect. 5.

[1] Which, according to some, would also account for the failure of the a priori character and the analyticity of logic; however, see Arenhart (2024) and Arenhart and da Cunha (2023) for further discussion.

[2] The same claim is also advanced by Erickson (2021), although we shall pursue it here in different lines.

2 AEL and Logical Realism

Let us begin by putting it again in explicit terms: AEL, as a general thesis, advances the claim that logic should have some kind of continuation with science, being able to join science in its successful achievements by using the same kind of methodology. As remarked, that claim concerning continuity is very flexible, given that the nature of such continuity may be explained in many distinct ways. All would be very unproblematic if there were a unified and consensual methodology of science, recognized by everyone and applied across the board. But things are not like that (see in particular the discussion about methodology in Bueno, 2017).

In order to put those issues in clearer terms, we begin, in this section, by presenting AEL with some of its nuances. Even though there is no uniformity in the distinct characterizations of AEL, some of the formulations found in the literature have basic common threads that will allow us to develop our point (for a criticism of the ambiguity in the characterization of AEL, see Rossberg and Shapiro 2021). Again, recall that one of the main targets of AEL is the alleged weaknesses of the traditional epistemology of logic, which should be overcome by advancing a science-based epistemology, focused on theory choice.

> Motivated by weaknesses with traditional accounts of logical epistemology, considerable attention has been paid recently to the view, known as AEL, that the subject matter and epistemology of logic may not be so different from that of the recognised sciences. One of the most prevalent claims made by advocates of AEL is that theory choice within logic is significantly similar to that within the sciences (Martin & Hjortland, 2021, p. 285).

The backdrop is to substitute current inappropriate epistemology (mainly thought to be a priori, recall) with the quite tested and approved epistemology of science. The focus lies, as the quote makes clear, primarily in theory choice, related to the claim that there are no—or there *should* be no—significant differences between science and logic on these matters. In this scenario, Martin and Hjortland (2021, p. 286) set up a *Methodological version of AEL* in the following terms: "Theory choice within logic is similar in important respects to that of the recognised sciences". So, according to this perspective, logical laws and inferences are justified—and perhaps come to be known—on the basis of theory choice, whose methodology is similar to the one employed by "recognised sciences".

The problem then is: how does theory choice work in the recognised sciences? In the methodological AEL camp, *abductivism* is often taken as the proper method for theory choice in the sciences, and logic should follow the same procedure (Priest, 2014, 2016; Hjortland, 2017; Martin, 2021). According to this view, logical propositions aren't justified directly on the basis of intuitions or semantic definitions (as the traditional accounts of logical epistemology would have it), but rather by being part of a logical theory that can be chosen as the best explanation of the available logical evidence. Roughly, one theory will be selected as the best explanation, mainly, by its ability to best accommodate the relevant data (which, remember, is supposed to be a posteriori), and also by taking into account a number of theoretical virtues, such as explanatory power, simplicity, avoidance of *ad hoc* elements, consistency (see Priest, 2016, p. 32).

That is not all to the methodological approximations, however. In the face of some problems with abductivism in logic,[3] and Martin and Hjortland (2022) have proposed a distinct formulation of AEL, arguing that logical theories are not chosen for their ability to fit the data, but rather by their ability to make successful predictions. According to Martin and Hjortland (2022, p. 148), this view, called *logical predictivism*, is the "most detailed and plausible version of methodological AEL available". In this view of logic, there are specifically logical phenomena that the logical theories try to explain, and the fruitfulness of these explanations is judged by the successful predictions (Martin & Hjortland, 2021, p. 288).

Despite their seeming differences, logical abductivism and logical predictivism have some common background assumptions. In both instances, the anti-exceptionalist aims to justify the correctness of a particular logical system by emphasizing a key feature of that system, one that sets it apart from rival systems that either lack it or possess it to a lesser extent. The abductivist wishes to conclude that a system is the correct one because it suits the evidence, while the predictivist wishes to argue for a system because it is taken to correctly predicts what is actually valid (remember also Ben Martin's quote at the beginning of this chapter).

What lies behind both approaches, we suggest, is basically a broadly realist conception of logic and of science.[4] The idea that there *is* a matter of fact about logical consequence, and that a common uniting feature behind the abductivist and the predictivist approaches is to argue that different theories try to describe or explain such facts. This is found in most discussions of what the data of logic could be (semantic data, or extra-linguistic data), and the view that the business of logic is description (and not, or not mainly, prescription). Among others, Graham Priest puts the point as follows, distinguishing a logical theory (*logica docens*) from the logical facts (*logica ens*):

> it is crucial to distinguish between logic as a theory (logic docens, with its canonical application), and what it is a theory of (logica ens). In the same way we must clearly distinguish between dynamics as a theory (e.g., Newtonian dynamics) and dynamics as what this is a theory of (e.g., the dynamics of the Earth). (Priest, 2014, p. 215).

And he further explains what *logica ens* consists of as follows:

> These are the facts of what follows from what—or better, to avoid any problems with talk of facts: the truths of the form 'that so and so follows from that such and such'. (Priest, 2014, p. 220).

According to the views proposed, then, logic, given that it simulates empirical science, is attempting to find out which among the many options available is the correct description of the notion of validity, or validity *simpliciter*. Both logic and science,

---

[3] See Hlobil (2020) and Arenhart (2024).

[4] For further critical discussion relating claims that abduction and scientific realism are part of the same piece, see van Fraassen (1980, Sect. 2.3) and the references therein.

on this view, are looking for the truth about a certain independently existing phenomenon, where truth is understood in a kind of correspondential sense (*viz.* discussions concerning 'the one true logic'). That is, this is an openly realist account of the goals and methodology of logic. Realism is a very difficult position to characterize, but, still, we can find the following commonality among the diversity:

> Amidst these differences, however, a general recipe for realism is widely shared: our best scientific theories give true or approximately true descriptions of observable and unobservable aspects of a mind-independent world. (Chakravartty, 2017).

Now, there is no question that it is perfectly fine to try to be a realist in logic. Depending on how the term is characterized, logical monists are realists of a sort, and perhaps most logical pluralists are realists too (i.e. when the characterization of these terms involves the idea of one or more than one correct or true description). But what cannot be forgotten is that realism is not the only game in town when it comes to the philosophical understanding of empirical science. As a result, it cannot be assumed without further ado that the *only* way to bring logic closer to science is by means of adhering to a realist description of the efforts of science. Of course, one may argue that realism is the *best* way of doing so, but then this should come up as a conclusion, not a premise. As Guay and Pradeu (2020, p. 1850) diagnosed, a similar attitude was taken in past discussions on metametaphysics concerning the relationship between science and metaphysics; as they called it, this was an "oversimplification" that "tended to reduce science-based metaphysical approaches to realism". Something similar, we claim, is happening to the efforts offered by AEL to bring logic closer to science.

We seek a parallel in the metaphysics of science, for we think there's a lesson to be learned from it. By avoiding a sort of restricted account of science, one opens up the scope of approaches to the understanding of science and, as a result, further possibilities that could help us to enlighten the understanding of logic are also now available. Why not, for instance, think that logic could be approached through the lights of distinct, not straightforward realistic, approaches to the philosophy of science? If one is on the search for approximating science and logic, one is also required to look for antirealist accounts of science. Of course, scientific antirealism comes in many distinct versions, and making one of those explicit would certainly open the way for distinct forms of AEL and antirealist approaches to logic. We are not claiming that they would all be equally interesting, but the fact that there are other options on the table makes the choice of a realist approach hardly obvious, to say the least.

One way of seeing how this could be done would be to adopt a form of constructive empiricist approach to logic—or, at least, to consider some of the features that constructive empiricism uses to approach science, and approach logic with the same tools. According to constructive empiricists, we should not believe in the truth of our scientific theories when it comes to the unobservable aspects of the description of reality; rather, given the evidence available, only belief in the empirical adequacy is warranted (van Fraassen, 1980, 1989). That would lead us to a different picture of the task of an epistemology for logic that connects with science. In fact, that is a dimension worth exploring a little bit, if by way of illustration only, of how an understanding of logic could benefit from distinct understandings of science.

According to the empiricist version of science, the existence of multiple theories, all equally compatible with the evidence, is a good reason for us to be agnostic about the literal truth of such theories (that is the moral of the famous argument from the underdetermination of the theories by the data, to be further explored in Sect. 3). A similar story could be provided in the case of logic by the fact that there are multiple logical frameworks, provided that they can give us the same important deductive results in many of the most important fronts (as some of them do with recapture strategies; see also our discussion in Sect. 4). So, just as distinct empirical theories compatible with the evidence would be telling us distinct ways the world could be, distinct systems of logic would be delivering for us distinct ways the logic behind our inferences could be (many-valued, constructive, inconsistency-tolerating, and so on), if each of such logical frameworks were true (van Fraassen, 1989).[5] What we would get from such a multiplicity of theories would be a better understanding of logical consequence, even if none of the candidates could be actually known to be 'the' true one.

For an illustration of how such an approach would benefit from actual evidence from science, we shall consider current debates concerning different levels of theory choice in quantum mechanics.

3 The Quantum Case

One of the crucial claims of AEL is related to the respectability, on epistemic grounds, that logic would derive from employing the same methodology of theory choice as the one employed by empirical sciences. As we have suggested in the previous section, the way the story is told in current literature, one is quickly led to a realist account of both science and logic, with the choice of a theory by abduction leading one to consider the chosen theory as a candidate to be the true theory.

Such a claim, however, rests on many unsettled issues. The most pressing one is that realism about science faces many difficult challenges, being diverse versions of underdetermination among the most pressing issues. Let's get specific about this by analyzing one of the favorite examples for philosophers of science: (non-relativistic) quantum mechanics. Quantum mechanics is the perfect medium for the realist/antirealist debate because it exemplifies very well what's at stake on both epistemic stances. On the realist side, it showcases undeniable empirical success. Quantum mechanics is outstandingly well-confirmed, so it enjoys all the empirical success a scientific theory can dream of. On the antirealist side, it showcases underdetermination, because quantum mechanics exemplifies such cases in so many distinct levels. Here we bring four such levels to the fore, respectively: dynamical,

[5] For alternative formulations of AEL not focusing on logical theory choice and not openly realistic, see for instance Arenhart (2022), Tajer (2022), and Commandeur (2023).

ontological, metaphysical, and logical levels. Let's briefly see how this plays out, and how an ensuing antirealist approach to logic may be motivated in this context too.

In familiar quantum-mechanical situations, one may describe the state of a quantum system (e.g. an electron) as a sum of its possible states. That is to say that, (suppose) if an electron e could assume the state of being located at A and thus assume the state-vector $|A\rangle_e$; and if it could *also* assume the state of being located at B and thus assume the state $|B\rangle_e$, the correct way of describing the electron in such an event is the sum $|A\rangle_e + |B\rangle_e$. This vector sum is called "superposition", and the field of quantum foundations spends lots of effort in spelling out whatever physical meaning it could amount to as it cannot be, for instance, meant that the electron is located in region A, nor in region B, nor in both A and B at the same time, and not in any of these places—which is the four logical possibilities we have of describing ordinary objects in our experience (Albert, 1992). However, it turns out that $|A\rangle_e + |B\rangle_e$ is the empirically adequate way of describing electrons in a wide array of experiments ranging from quantum optics to Nobel-prize winning quantum correlation experiments (Lewis, 2016). Something must be going on there, and quantum foundations is the field that (among other things) is supposed to explain precisely what is it. The usual way of presenting what's wrong with the above picture goes something like this, in an either-or situation that started out as a dilemma (Bell, 1989), evolved into a trilemma (Maudlin, 1995) and now it's better described as polylemmas (Muller, 2023):

- The superposition is couched in the wrong terms. If the dynamics led us into a superposition situation, it should be plainly wrong as we never get to observe superposition—not in the physics lab, not in daily life. There must be *another* kind of dynamics that gives us a determinate position (to stick with the example), and this dynamics is often called the "collapse". So there's a dynamics for superposition, and another one to get rid of it. This is often called standard quantum mechanics, or simply "quantum mechanics".
- The superposition is couched in incomplete terms, qua epistemic, so that the position of the electron is always determinate albeit we don't have epistemic access to it. So the dynamics is often said to be one of hidden variables, or hidden parameters, that specifies the determinate electron's position at all times. The way to go for it is e.g. Bohmian mechanics.
- The superposition is read literally: both states describe what is actually going on, albeit not in the same world. No need to add hidden variables or other dynamics: the dynamical law that describes the physical situation as a superposition is everything and it is right. This is the Everettian quantum mechanics way to go.

This is how far science goes, delivering us the so-called 'quantum interpretations'. What is illustrated is the first level of underdetermination (of theory by evidence), and there's no *scientific* reason to choose one among the options offered. This is to say that physics alone does not determine which is the right way to go. Notice that this does not mean that absolutely everything goes, as the interpretations must be empirically adequate to current evidence. To stress our methodological point, however, the following must be clear: QM does not determine its interpretation.

In the remainder, we'll stick with the first case in the textbook story mentioned before, the one of standard quantum mechanics. Recall that no *scientific* reason compels one to do so; we're picking this one for the sake of building our argument.

So let's keep digging. The ontological level deals with what exists according to each interpretation. To reiterate, we are dealing with a single interpretation, and there's no consensus here as well. For instance, one might put forth the idea according to which the furniture of the world is composed of (among other things) particles in three-dimensional space and human consciousness that cause the collapse dynamics (de Barros & Oas, 2017); alternatively, one can *also* point out that there are no particles or minds, and what exists in the fundamental level is a high-dimensional field called the wave function (Ney, 2021). Surely there are more alternatives than that, but two will suffice. This is the second level of underdetermination, and—just as it occurred with the first level—there's no *scientific* reason to choose among them. It doesn't matter which one we pick for us to keep digging beyond this point, as long as we keep committed to our first choice, *viz.* standard quantum mechanics.

Enter the metaphysical level. Due to the indistinguishability postulate, the permutation of quantum objects does not give rise to a new physical situation (Huggett & Imbo, 2009). Given any two physical systems, say, $|\psi\rangle$ and $|\phi\rangle$, their permutation of their tensor product $\otimes$ results in the equality $|\psi\rangle \otimes |\phi\rangle = |\phi\rangle \otimes |\psi\rangle$. This is often contrasted with the so-called "classical" case, which can be illustrated with a statistical example. Suppose one wants to arrange two particles, 1 and 2 inside two boxes, A and B; the four classical possibilities (C_1–C_4) to do so are:

C_1. A^1A^2: Both particles are inside box A;
C_2. B^1B^2: Both particles are in box B
C_3. A^1B^2: Particle 1 is inside box A, and 2 in B;
C_4. B^1A^2: Particle 1 is inside box B, and 2 in A.

The statistical weight for each one of these possibilities is $\frac{1}{4}$, and this is called the Maxwell–Boltzmann statistics.

So far, so good, but QM disrupts that with the indistinguishability postulate. Notice that the third and fourth cases differ only in particle permutation, so the statistics cannot generate a new physical situation. Here's what happens, then, with the quantum possibilities (Q_1–Q_3):

Q_1. $A^{\bullet\bullet}B^{\circ\circ}$: Both indistinguishable particles are inside box A;
Q_2. $A^{\circ\circ}B^{\bullet\bullet}$: Both indistinguishable particles are inside box B;
Q_3. $A^{\bullet\circ}B^{\bullet\circ}$: One indistinguishable particles is inside each box, A and B.

Q_1–Q_3 have the same statistical weight. Notice how Q_3 replaces cases C_3 and C_4. This is called the Bose–Einstein statistics.

The above-mentioned change in the statistics for the classical and quantum cases is often explained by a change in the *metaphysics* of quantum objects. In the traditional way of thinking about this situation, it is thought that, whereas classical objects are endowed with *individuality*, quantum objects are not. This would explain why C_3 and C_4 are different physical situations, and why they all collapsed into Q_3: quantum objects have no individuality, so it doesn't matter if one exchanges them. Quantum objects, in this view, are non-individuals (Krause et al., 2022). Whereas classical objects can be cashed out in terms of haecceities (the non-qualitative property of "being identical with itself", quantum objects should be understood by their lack of haecceity. This is the Received View of quantum (non-)individuality (French & Krause, 2006; Arenhart, 2017). Quantum objects, however, can *also* be cashed out in haecceistic terms—just as their classical cousins—if one wishes to do so (French, 2019). One might simply point out that haecceities are spelt out in substantialist terms, or simply posited as primitive (Morganti, 2015).

The methodological point that we wish to stress with all this is the following: *none of the above-mentioned metaphysical views for quantum objects can be settled by QM itself.* This is underdetermination level three: the "underdetermination of metaphysics by physics" or simply "metaphysical underdetermination" (French & Krause, 2006). The individuality/non-individuality profiles may always be added as a metaphysical maneuver to further interpret quantum objects *qua* objects without changing physical/empirical aspects of quantum mechanics. Because of that, the physical/empirical domain cannot tell which of such metaphysical 'packages' is more appropriate/true. At the end of the day, metaphysics of science is still metaphysics—and the fact that it is applied to science doesn't make much progress in terms of theory choice (Arroyo & Arenhart, 2022).

On to logical underdetermination at level four. Take the case for quantum statistics, specifically Q_3. The particles inside each box, A and B, are *absolutely indistinguishable*. One way of interpreting it is to state that quantum objects lack the notion of "sameness" (to use the words of Schrödinger, 1998, p. 197), and another way is stating that they don't.

In logical terms, it has been suggested that it all comes down to whether or not the reflexivity of identity—understood in the terms of $\forall x(x = x)$—is being violated by quantum mechanics. If not, classical logic is everything one needs in this matter, but then one must do some tricks to account for absolute indistinguishability (see French and Krause 2006; Krause and Arenhart 2018 and Bigaj, 2022 for discussion). If reflexivity of identity is being violated by Q_3, then one must jump right into some non-classical system. There's one tailored made for this case: Schrödinger logics (da Costa & Krause, 1994) and its set-theoretical level through the use of quasi-sets (Krause, 1992; French, 2019). Roughly speaking, in such non-reflexive logics the expression $x = x$ is a well-formed formula only in specific cases, *viz.*, cases in which the objects in question are not quantum objects, but so-called "everyday" ones (such as tables, chairs, and so on). For quantum objects, identity ("=") doesn't apply; for everyday objects it does. For the former, quasi-set theory offers a weaker relation, the one of indiscernibility ("$\equiv$").

What is important for us now, however, are not the details of how such theories would work, but rather, that non-reflexive logics, in the form of quasi-set theory, seem to encompass a metaphysics of non-individuality. In this sense, they rival classical logic approaches to identity, which, it is claimed, codify individuality right from the start (see French and Krause 2006, Chap. 6). The trouble is that with the underdetermination of the metaphysics by the physics, we are stuck also with no additional evidence for the choice of the appropriate logic, classic or non-reflexive. In fact, both classical set theory and quasi-set theory can provide for the mathematical frameworks needed to develop quantum mechanics, so that there is no mathematical or physical fact of the matter that could help us to choose one of the logics.[6] The result is deep underdetermination also for logic.[7]

Now, notice that all these kinds of underdetermination happen in the *same domain* of investigation. It is not as if we're having trouble unifying the dynamics/ontology/metaphysics/logics of e.g. non-relativistic quantum mechanics and special relativity, nor quantum mechanics and pediatrics. A pluralist account may handle that case more quickly, stating that each domain has it own specifics (see, for example, Bueno, 2010; Ruetsche, 2015; Bueno, 2017). Bueno (2010), for instance, argues that we should not bother deciding between the logics of quantum mechanics (the so-called "quantum logic"[8] presented by Birkhoff and von Neumann 1936) and logics of other domains; Ruetsche (2015) has a similar move by stating that we should peacefully accept the plurality of theories (specifically dynamics and ontology) for different domains. The cases we've brought to the table are even more radical, *viz.*, of quantum mechanics itself having no means whatsoever to decide these matters on purely quantum-theoretical grounds.

As we have seen, the appeal to continuity with science (whatever that means exactly) was not enough to settle the four above-mentioned levels of underdetermination. Physics is deathly silent on such matters, being compatible with every single one of the aforementioned possibilities. This means, among many other things, that quantum mechanics has no clear method for theory choice. So when it comes to the matter of choosing a correct logical theory concerning some aspect of "logical facts", quantum mechanics is not your ally. The moral of this example is what went wrong: AEL shouldn't be so quick in hoping that science is always teaching us clear lessons about the true theory to be chosen. Underdetermination is a widespread phenomenon, and the antirealist makes a lot of it.[9]

[6] See Domenech et al. (2008) for a development of Fock spaces in quasi-set theory.

[7] See Arenhart and Arroyo (2023) for discussion on these levels of underdetermination.

[8] We should mention in passing that quantum logic, like quasi-set theory, was tailored-made for the quantum case; unlike quasi-set theory, however, quantum logic doesn't address specific questions of identity and indiscernibility.

[9] This is not to say that underdetermination is a knockdown argument against scientific realism, though. Alternatives are available. To mention a few, it has been proposed that: we should leave quantum mechanics out of the scientific realist picture (Hoefer, 2020); we should explicitly use extra-empirical virtues to theory choice in these cases (Callender, 2020); that there are commonalities between underdetermined options that must be sought (French, 2014; Fraser & Vickers, 2025). All of them compromise the scientific realist starting goal a little bit, but that's not why we brought up underdetermination arguments. Scientific realist's attitude towards underdetermination be as it may, the point we wish to stress is that theory choice in science is far from being unproblematic.

4 The Lessons for AEL

As we have illustrated with the case of quantum mechanics, there are at least some reasons to believe that an antirealist account of science should also be considered when issues related to the continuity between logic and science are taken into account. As a result, those looking for inspiration in science could also attempt to develop accounts of the methodology of logic that are not straightforwardly realist, and check what are the implications of giving up the kind of realist approach that current anti-exceptionalist views have offered. Abandoning the unquestioned view that the aim of an epistemology for logic is to deliver the means to choose the true theory/theories is already a nice pointer towards a more productive engagement with science, according to antirealist lights. Here, we list some of the additional benefits that such an exploration could enjoy.

- We can overcome the uncritical view that there is a commonly accepted and successful description of scientific methodology, and that philosophers of logic are to blame for not having attempted to copy such a methodology before. The fact is that science itself is not a reign of consensus where the epistemology and metaphysics are worn on its sleeves; rather, consensus about the most important points appears only at the surface. Science is a very intricate human enterprise, and although its success is undeniable, it is also undeniable that the belief that it can deliver many instructive lessons as to how to achieve a similar level of results is not easily implemented (as the discussion on the metaphysics of science in the previous section witnesses).
- Once one gives up the idea that the aim of logical inquiry is the search for the one true theory, the focus can shift to our *understanding* of logic as it is, not as how philosophers would like it to be pursued. The fact is that logic is already a science, highly investigated and developed not only in philosophy departments, but also in mathematics and computer science departments, with its own research methods and research programs well established. Attempting to making the search for the one true logic as the major task for the epistemology of logic is at odds with current logical practice, at least when logicians are doing logic as a kind of investigation into the properties of logical systems. As Franks (2015) has put it:

  > One should warm up to the trend of identifying logic with the specialized scientific study of the relationships among various systems and their properties. This is, after all, how logicians use the word. Our preference to ignore questions about a logic's correctness stems not only from an interest in exploring the properties of possible logical systems in full generality but also from an appreciation, fostered by the study of logic, that no one such system can have all the properties that might be useful and interesting. (Franks, 2015, p. 163).

  By shifting to an antirealist way of pursuing the investigation of logic, instead of centering the whole epistemology in the search for the one true theory, we seem to end up much closer to current logical practice.[10] In the practice of logicians, questions of the correction of a system are of minor importance. Closely representing the practice of a science is seen in philosophy of science as a virtue of a philosophical account, while distance from practice is a negative point (as the debate between logical positivists and more practice-oriented approaches witness); the epistemology of logic could also learn this particular lesson.

- Given the important role that classical logic plays in empirical science and classical mathematics, one could be led to think that this is the logic to be chosen as the true one (see the discussion in Hjortland 2017, 2019; Bueno 2017). However, to recover a topic mentioned before, discussions related to the recapture of classical logic by non-classical theories in scenarios where classical logic is actually doing an important job should illustrate that there seems to be no fact of the matter that could help us to choose a logic for empirical science in general. As illustrated by Hjortland, in scenarios where one is willing to investigate the use of paracomplete logic in order to ground truth predicates and avoid semantic paradoxes, one needs not abandon classical mathematics:

  > But the theories nonetheless preserve classical mathematics. In model-theoretic terms, the fixed-point models of paracomplete theories are classical for the arithmetic language: no arithmetic sentence receives a nonclassical truth value. Importantly, the logical concepts of the paracomplete theories behave classically for sentences with classical truth values. As a result, the classical principles such the law of excluded middle are recaptured for arithmetic sentences and arguments. Of course, the presence of the truth predicate will complicate matters, but for good reason. The theory would be inconsistent if non-arithmetic sentences behaved classically. For the record, it is not only paracomplete theories that have this recapture property, but a number of nonclassical theories of truth (e.g. paraconsistent theories, substructural theories). (Hjortland, 2019, p. 272).

  The result is that if science is to provide part of our evidence, as an anti-exceptionalist seems to believe, we seem to have to rely on considerations that are clearly not truth conducive (such as simplicity, familiarity, elegance, and so on) to guide our choices, in case we believe that a logic must be chosen. That is, of course, not a problem for an antirealist, but it clearly is an obstacle for those willing to apply abduction as a source of discovery of the one true logic.
- By rejecting the idea that there must be some kind of true/correct logic, the one that is responsive to the logical facts, we unburden our metaphysics. In fact, this does away with the very idea that there must be an (inaccessible) realm where logical consequence, validity itself, or *logica ens* resides. Such an idea seems to be quite contrary to the ones guiding the investigation of logic and its diverse systems. Without requiring inaccessible logical facts, one brings logic closer to metaphysically deflationary accounts of science, and seems to account for some important views concerning the very nature of logical consequence. As Peter Smith has put it:

[10] That closeness to the practice is a virtue of epistemology for logic is also recognized by Martin (2021).

> If you think that there is [such a notion], start asking yourself questions like this. Is the intuitive notion of consequence constrained by considerations of relevance?—do ex falso quodlibet inferences commit a fallacy of relevance? When can you suppress necessarily true premises and still have an inference which is intuitively valid? What about the inference 'The cup contains some water; so it contains some H_2O molecules'? That necessarily preserves truth (on Kripkean assumptions): but is it valid in the intuitive sense?—if not, just why not? (Smith, 2011, p. 29).

All of those enigmas disappear with the antirealist approach, counting as another benefit of the view.

- Consider the result of the realist approach to science and AEL regarding the issue of how many systems may be chosen as the correct one. Even though Hjortland (2017) concedes as unattainable the idea that singling out a specific logic could be the final outcome of an abduction process, he still holds on to the idea that distinct logics may capture distinct regions of validity and this generates a pluralism of some sort. Priest (2014) puts the issue in more neutral terms regarding the debate about monism versus pluralism:

> I observe that this procedure does not prejudice the question of logical monism vs logical pluralism. If there is "one true logic" one's best appraisal of what this is determined in the way I have indicated. If there are different logics for different topics, each of these is determined in the same way. Whether one single logic is better than many, is a "meta-issue", and is itself to be determined by similar considerations of rational theory-choice. (Priest, 2014, p. 216).

As a result of giving up the belief that some system may be true, the debate between pluralism and monism in the philosophy of logic ends up deflated. This is certainly a topic to be explored, and the antirealist emphasis on understanding in opposition to truth clearly finds a place here as well.

5 Conclusion

AEL is notoriously difficult to characterize, given the flexibility with which it has been characterized in the literature. However, taking into account the methodological agenda of AEL, which has been present in most of the discussion about AEL, there is an underlying assumption concerning a new epistemology related to a methodology for theory choice, and also, the assumption that we will succeed in such a search for a new epistemology—ideally better than the traditional one—if we can choose the correct theory/theories among the many options available.

In this paper, we have sought to uncover these underlying assumptions and, given the anti-exceptionalist willingness to approach science, we have suggested that there are alternative ways to understand science that may be also promising for the anti-exceptionalist. That is, given that the lessons of science itself are not completely clear, the bridge between logic and empirical science is also not completely obvious. We have suggested that antirealist approaches to science also offer appropriate ways to understand science and logic, so that some of the difficulties plaguing current AEL may be understood as a result of using realist approaches to science. Although we have not explicitly defended any antirealist approach in particular, offering such additional possibilities is also part of the tools the antirealist has to understand science. We hope this sheds some light on further avenues that may be explored concerning AEL.

Acknowledgements The order of authorship is alphabetical. The authors would like to thank: Evelyn Erickson for discussions on some of the topics explored in this paper; Sofia Abelha Meirelles for reading and commenting on an earlier version of this paper; the Philosophy of Logic reading group participants, hosted by the GFILL/CNPq Research Group and the Graduate Program in Philosophy of the Federal University of Maranhão. We also thank Sanderson Molick for the kind invitation for us to contribute to this volume with many authors we admire. Raoni Arroyo was supported by grants #446478/2024-5 and #315067/2025-0, Conselho Nacional de Desenvolvimento Científico e Tecnológico (CNPq), call CNPq/MCTI/FNDCT #21/2024. The earlier versions of this chapter were developed during research visit to the Department of Philosophy, Communication and Performing Arts of the Universitá degli Studi Roma Tre, Italy, supported by grant #2022/15992-8, São Paulo Research Foundation (FAPESP). Jonas R. B. Arenhart is partially supported by CNPq.

References

Albert, D. Z. (1992). *Quantum mechanics and experience*. Cambridge: Harvard University Press.

Arenhart, J. R. B. (2017). The received view on quantum non-individuality: Formal and metaphysical analysis. *Synthese, 194*, 1323–1347. https://doi.org/10.1007/s11229-015-0997-5

Arenhart, J. R. B. (2022). Logical anti-exceptionalism meets the "Logic-as-models" approach. *Theoria, 88*(6), 1211–1227. https://doi.org/10.1111/theo.12436

Arenhart, J. R. B. (2024). Abductivism as a new epistemology for logic? *Erkenntnis, 89*, 2397–2415. https://doi.org/10.1007/s10670-022-00634-7

Arenhart, J. R. B., & Arroyo, R. (2023). The roads to non-individuals. In J. R. B. Arenhart, & R. Arroyo (Eds.), *Non-reflexive logics, non-individuals, and the philosophy of quantum mechanics: Essays in honour of the philosophy of Décio Krause*. Synthese Library, vol. 476 (pp. 79–100). Berlin: Springer. https://doi.org/10.1007/978-3-031-31840-5

Arenhart, J. R. B., & da Cunha, I. F. (2024). Carnapian lessons for anti-exceptionalism about logic. *History and Philosophy of Logic, 44*(1), 54–65. https://doi.org/10.1080/01445340.2022.2162327

Arroyo, R., & Arenhart, J. R. B. (2022). The epistemic value of metaphysics. *Synthese, 200*(337). https://doi.org/10.1007/s11229-022-03833-5

Bell, J. S. (1989). Are there quantum jumps? In C. W. Kilmister (Ed.), *Schrödinger: Centenary celebration of a polymath* (pp. 41–52). Cambridge: Cambridge University Press.

Bigaj, T. (2022). *Identity and indiscernibility in quantum mechanics*. Palgrave Macmillan.

Birkhoff, G., & von Neumann, J. (1936). The logic of quantum mechanics. *Annals of Mathematics* 823–843.

Bueno, O. (2010). Is logic a priori? *The Harvard Review of Philosophy, 17*(1), 105–117.

Bueno, O. (2017). Scientific pluralism, consistency preservation, and inconsistency toleration. *Humana Mente, 10*(32), 229–245.

Callender, C. (2020). Can we quarantine the quantum blight? In S. French & J. Saatsi (Eds.), *Scientific realism and the quantum* (pp. 55–77). New York: Oxford University Press.

Chakravartty, A. (2017). Scientific realism. In E. N. Zalta (Ed.), *The Stanford Encyclopedia of Philosophy*, summer 2017 edn. Metaphysics Research Lab, Stanford University. https://plato.stanford.edu/archives/sum2017/entries/scientific-realism/

Commandeur, L. (2023). Logical instrumentalism and anti-exceptionalism about logic. *Erkenntnis*. https://doi.org/10.1007/s10670-023-00752-w
da Costa, N. C. A., & Krause, D. (1994). Schrödinger logics. *Studia Logica, 53*(4), 533–550.
de Barros, J. A., & Oas, G. (2017). Can we falsify the consciousness-causes-collapse hypothesis in quantum mechanics? *Foundations of Physics, 47*(10), 1294–1308.
Domenech, G., Holik, F., & Krause, D. (2008). Q-spaces and the foundations of quantum mechanics. *Foundations of Physics, 38*(11), 969–994. https://doi.org/10.1007/s10701-008-9246-9
Erickson, E. F. (2021). On the metaphysics of (Epistemological) logical anti-exceptionalism. *Principia: An International Journal of Epistemology, 25*(1). https://doi.org/10.5007/1808-1711.2021.e80109
Franks, C. (2015). Logical nihilism. In Å. Hirvonen, J. Kontinen, R. Kossak, & A. Villaveces (Eds.), *Logic without borders: Essays on set theory, model theory, philosophical logic and philosophy of mathematics* (pp. 147–166). De Gruyter.
Fraser, J. D., & Vickers, P. (2025). Knowledge of the quantum domain: An overlap strategy. *British Journal for the Philosophy of Science, 76*(3). https://doi.org/10.1086/721635
French, S. (2019). Identity and individuality in quantum theory. In E. N. Zalta (Ed.), *The Stanford Encyclopedia of Philosophy*, winter 2019 edn. Metaphysics Research Lab, Stanford University. https://plato.stanford.edu/archives/win2019/entries/qt-idind/
French, S. (2014). *The structure of the world: Metaphysics and representation*. Oxford: Oxford University Press.
French, S., & Krause, D. (2006). *Identity in physics: A historical, philosophical, and formal analysis*. Oxford: Oxford University Press.
Guay, A., & Pradeu, T. (2020). Right out of the box: How to situate metaphysics of science in relation to other metaphysical approaches. *Synthese, 197*(5), 1847–1866.
Haack, S. (1978). *Philosophy of logics*. London and New York: Cambridge University Press.
Hjortland, O. T. (2019). What counts as evidence for a logical theory? *Australasian Journal of Logic, 16*(7), 250–282. https://doi.org/10.26686/ajl.v16i7.5912
Hjortland, O. T. (2017). Anti-exceptionalism about logic. *Philosophical Studies, 174*, 631–658.
Hlobil, U. (2020). Limits of Abductivism about logic. *Philosophy and Phenomenological Research, 103*(2), 320–340. https://doi.org/10.1111/phpr.12707
Hoefer, C. (2020). Scientific realism without the quantum. In S. French & J. Saatsi (Eds.), *Scientific realism and the quantum* (pp. 19–34). New York: Oxford University Press.
Huggett, N., & Imbo, T. (2009). Indistinguishability. In D. Greenberger, K. Hentschel, & F. Weinert (Eds.), *Compendium of quantum physics: concepts, experiments, history and philosophy* (pp. 311–317). Dordrecht: Springer.
Krause, D., & Arenhart, J. R. B. (2018). Quantum non-individuality: Background concepts and possibilities. In S. Wuppuluri, & F. A. Doria (Eds.), *The map and the territory: Exploring the foundations of science, thought and reality*. The frontiers collection (pp. 281–305). Berlin: Springer.
Krause, D. (1992). On a quasi-set theory. *Notre Dame Journal of Formal Logic, 33*(3), 402–411.
Krause, D., Arenhart, J. R. B., & Bueno, O. (2022). The non-individuals interpretation of quantum mechanics. In O. Freire Jr. (Ed.), *The oxford handbook of the history of quantum interpretations* (pp. 1135–1154). New York: Oxford University Press.
Lewis, P. J. (2016). *Quantum ontology: A guide to the metaphysics of quantum mechanics*. New York: Oxford University Press.
Martin, B. (2021). Identifying logical evidence. *Synthese, 198*, 9069–9095.
Martin, B., & Hjortland, O. T. (2021). Logical predictivism. *Journal of Philosophical Logic, 50*(2), 285–318. https://doi.org/10.1007/s10992-020-09566-5
Martin, B., & Hjortland, O. T. (2022). Anti-exceptionalism about logic as tradition rejection. *Synthese, 200*(2), 1–33. https://doi.org/10.1007/s11229-022-03653-7
Maudlin, T. (1995). Three measurement problems. *Topoi, 14*(1), 7–15.
Morganti, M. (2015). The metaphysics of individuality and the sciences. In A. Guay & T. Pradeu (Eds.), *Individuals across the sciences* (pp. 273–294). Oxford: Oxford University Press.

Muller, F. A. (2023). Six measurement problems of quantum mechanics. In J. R. B. Arenhart, & R. Arroyo (Eds.), *Non-reflexive logics, non-individuals, and the philosophy of quantum mechanics: Essays in honour of the philosophy of décio krause*. Synthese library, vol. 476 (pp. 225–259). Berlin: Springer.
Ney, A. (2021). *The world in the wave function*. Oxford: Oxford University Press.
Priest, G. (2014). Revising logic. In P. Rush (Ed.), *The metaphysics of logic* (pp. 211–223). Cambridge: Cambridge University Press.
Priest, G. (2016). Logical disputes and the a priori. *Princípios, 24*(40), 29–57.
Rossberg, M., & Shapiro, S. (2021). Logic and science: Science and logic. *Synthese, 199*(3–4), 6429–6454. https://doi.org/10.1007/s11229-021-03076-w
Ruetsche, L. (2015). The shaky game+ 25, or: on locavoracity. *Synthese, 192*(11), 3425–3442.
Schrödinger, E. (1998). What is an elementary particle? In E. Castellani (Ed.), *Interpreting bodies: classical and quantum objects in modern physics* (pp. 197–210). Princeton: Princeton University Press.
Smith, P. (2011). Squeezing arguments. *Analysis, 71*(1), 22–30. https://doi.org/10.1093/analys/anq120
Tajer, D. (2022). Anti-exceptionalism and methodological pluralism in logic. *Synthese, 200*(3), 1–21. https://doi.org/10.1007/s11229-022-03698-8
van Fraassen, B. C. (1980). *The scientific image*. Oxford: Oxford University Press.
van Fraassen, B. C. (1989). *Laws and symmetry*. Oxford: Oxford University Press.

How to Revise Logic: On Reflective Equilibrium, Upward Normative Pressures and a Posteriori Revisions

Marcos Silva

Abstract The paper addresses the problem of justifying fundamental logical principles, asking how they can be assessed without falling into vicious circularity or infinite regress, given that any form of rational justification appears to presuppose them. It argues that logic should be understood as a normative science, analogous to disciplines such as ethics and politics, a view also endorsed by Prawitz. Drawing on insights from semantic inferentialism and logical expressivism, particularly in Brandom, the paper proposes a neopragmatist method for the revision of logic based on reflective equilibrium between general theoretical considerations and particular inferences. Within this framework, the notion of normative bidirectionality is developed, with special emphasis on what is termed "upward normative pressure," which captures the dynamic and a posteriori character of the revision of logical principles. This notion helps to explain the emergence of non-classical logics by analogy with revisions in political and ethical normative systems. Rather than defending reflective equilibrium against its critics, the paper aims to show that it can serve as a viable method for revising logic in a context where the normativity of logic is grounded in the normativity of natural language, in line with the inferentialist perspective.

1 Introduction

Brouwer's intuitionist program, as presented in his historical Ph.D. thesis and paper *The Unreliability of the Logical Principles* (1908), shows a genuine philosophical motivation to revise logical principles, not just for engineering innovations in the introduction of formal calculi. Brouwer argues that we must revoke the universal and unrestricted validity of the Principle of the Excluded Middle and reject indirect proofs. Thus, in Heyting's intuitionist systems (Heyting, 1928), which develop Brouwer's thinking, an inference rule is valid just if a construction can be found that makes true the statement that is obtained by applying the rule. On this view, the principles of logic do not have to preserve, as in classical logic, mind-independent

M. Silva (✉)
Federal University of Pernambuco, Recife, Pernambuco, Brazil
e-mail: marcossilvarj@gmail.com

S. Molick (ed.), *Demarcating Logic and Science*, Studies in Applied Philosophy, Epistemology and Rational Ethics 80, https://doi.org/10.1007/978-3-032-25674-4_6

truth, but must be (mentally) constructible. This results in various principles of classical logic, most notably the Principle of the Excluded Middle, being rejected as insufficiently grounded.

The unreasonableness of the classic Principle of Explosion also presents compelling grounds to revise logical principles inasmuch as it should not be possible to derive every possible sentence from a contradiction (Da Costa, 1958, 1959, 1974; Marcos, 2004, 2010). In classical logic, it suffices to have a contradiction to lead the relation of logical consequence to trivialization. In fact, the problem is not the existence of contradictions, but the occurrence of an explosive logical consequence relation. Trivialization must be distinguished from contradiction (Marcos, 2004, 2010). In some rational discussions or in some important domains of our lives, when we are dealing with beliefs and information, we encounter contradictions yet continue to reason while resisting the unrealistic demand to commit to any other belief, especially peculiar ones (Carnielli & Rodrigues, 2015, 2016, 2017, 2019a, b). Furthermore, it is worth noting that some theories in science can be inconsistent without being trivial (Priest et al., 2015).

These philosophical objections to the nature and scope of classical logical orthodoxy have paved the way for the emergence of intuitionist and paraconsistent alternatives to classical logic. These families of logic have important philosophical and technical implications for mathematics and computation. I take the examples above concerning the emergence of both intuitionist logics and paraconsistent logics as illustrative of what I mean by revising logical principles from a philosophical point of view.

Moreover, we should take the current plurality of non-classical logic as a serious philosophical question, that is, as something which demands for philosophical explanation and requires that we reconsider the very role and purpose of logic. Two extreme views have prevented serious engagement with the possibility of a plurality of real alternative logic. On the one hand, a vision may take logic as some kind of metaphysical investigation or enterprise that could offer us sophisticated intellectual tools to discover independent, abstract, and a priori structures of our cognitive activities outside our minds. On the other hand, another view, a kind of conventionalism, may maintain that logic is basically a game, empty of content, governed by ad hoc and arbitrary decisions and rules. I hold that we can avoid both way-outs using lessons drawn from semantic inferentialism and logical expressivism as defended in Brandom's (1994, 2001) and Peregrin's work (2014). The emphasis in revising logic should be put on (i) how the meaning of logical vocabulary is derivative of our knowing how to operate regular inferential patterns that articulate the meaning of our current concepts, which means semantic inferentialism, and (ii) on the expressivist idea that we use logical vocabulary to make explicit inferential transitions that we already master implicitly in our daily discourse exchanges.[1]

---

[1] In doing so, I think there is no tension that needs to be clarified between the line that reflective equilibrium does not assume the existence of a metaphysical basis nor does presuppose a pure conventionalist approach to logic. The current theory of logical conventionalism, particularly defended by Warren (2020), defends that the logical consequence relation of a language is explained by the

In this paper, we combine inferentialism and expressivism with the notions of reflective equilibrium and normative bidirectionality to defend a neopragmatist and anti-foundationalist view of revising logic. We argue that a particular notion of reflective equilibrium, informed by semantic inferentialism and logical expressivism, offers an adequate conceptual platform for the discussion about the philosophical foundations of the revision of logic. Prima facie, I am not interested here in reflective equilibrium as a methodology to do real revision. I am interested in a philosophical justification to unify heterogeneous philosophical motivations for revising logical principles, such as the one from intuitionism and one that gives rise to paraconsistent logics. Moreover, we argue that the notion of reflective equilibrium is rich enough to allow and express the dynamics of the a posteriori revision process of some non-classical logics.[2] The following points are going to be discussed and articulated: the normative character of logic grounded on semantic inferentialism and logical expressivism; the dynamical and a posteriori aspect of revising logic; the normative bidirectionality and then some upward and downward normative pressures on revision. We contend that, while the downward normative pressure is ubiquitous when we, for instance, teach logic to undergrads (i.e. to learn how to use general principles to revise local applications of inferences), the introduced notion of upward normative pressure motivates some of the most interesting cases of a posteriori revision or restriction of logical principles (that is to pressure changes of general principles by the gained relevance of some local applications of inferences).[3] Further, upward normative pressures show how to see the rise of non-classical logic in analogy of how political and ethical systems might be revised as well. It is also noteworthy to say that it is not my aim to defend Reflective Equilibrium against detractors in this paper. My contribution here is about showing how to use Reflective Equilibrium as a method to revise logic in a horizon where the normativity of logic is due to the normativity of natural language, as in a usual inferentialist point of view.

---

linguistic conventions of that language. My proposal of looking at the roots of logic in our linguistic inferential patterns transcends mere logical conventionalism because I am favoring logical expressivism and semantic inferentialism. What follows is not a conventionalist proposal as a Carnapian one, but a Brandomian one, as we do not choose the natural language and culture we are introduced to in early ages. Our first natural language is our first Neurath's boat as we will see in Sect. 3.

[2] One might ask: How can a method express something if a method is the means through which something is achieved? I do not agree with this way of thinking. A method in philosophy is not just that. A method also expresses a philosophical view and shows one's philosophical assumptions. Further, a method can be materially adequate or inadequate. For instance, a Carnapian conventionalist view seems to be inadequate to capture what goes on historically in the revision of logics. Logicians are not at liberty to choose. They have natural and normative pressures, as institutional ones. I do not have the ambition to offer a unique method to revise logics. I am a pluralist concerning methodologies. Many methods and the combination of them might do the work. How to combine different methods to revise logic is beyond the scope of this contribution.

[3] We will discuss some examples in the fourth section of this paper. The restriction of the Principle of Excluded Middle and the Principle of Explosion are interesting for me because I start the paper with Browuer's view and with the Brazilian paraconsistent tradition. But our neopragmatist view on reflective equilibrium guiding revision of logics might be generalizable and used in other nonclassical logics not just in intuitionism and paraconsistency.

The article is divided into four sections. The first presents a general epistemological problem concerning the philosophy of logic. The second investigates the revision of logic and its intrinsically normative character using inferentialist and expressivist insights. In the third section, we examine reflective equilibrium and its relation to a pragmatist and anti-foundationalist view of revising logic. Finally, we develop our pragmatist view by introducing the notion of normative bidirectionality and argue that upward normative pressure can yield an interesting model to express the dynamics of a posteriori revision of logical principles.

2 On General Epistemological Problems Concerning the Philosophy of Logic

The current great plurality of logics gives rise to a general epistemological problem, one that can be applied in the discussion concerning the nature of logic and the very possibility of justifying and revising its rules and most basic principles. One may question whether it is possible (and how to) choose the most basic rules of a system. After all, how could our logical principles be justified if the very possibility of rational justification presupposes them? In fact, a cluster of epistemological questions can be envisioned here. How should we argue about a basic set of rules of inference and logical principles? Which rational arguments can be used to convince an interlocutor that a basic set of rules is the correct one if any argument must be, from the beginning, based on a set of accepted inferential rules? These questions concern the very nature of logic and rationality and can be applied to its justification, always being possible to question how reason can be used to substantiate and eventually revise the most basic principles of reason itself without circularity or infinite regress. Is there a rational way to convince someone of something as fundamental as the legitimacy of a principle of logic?[4]

According to Bueno and Colyvan (2004), the emergence of non-classical logics motivates the revision of many philosophically attractive features of logic and its centrality for rationality and our daily lives. In fact, philosophical perspectives that emphasize special predicates for logic regularly relate it to necessity, universality, apriority and, somehow, also to eternity. However, all these features have been severely challenged during the last decades in philosophical research. About these difficulties Bueno and Colyvan (2004), for instance, state:

> One of the reasons that philosophy of logic is such a difficult enterprise is that, in order to advance debates in this area, we require the very thing we are studying: logic. This difficulty is especially acute when engaging in the business of theory choice in logic. After all, in order to decide between two logical theories, we need to put forward evidence and arguments for each. This evidence and these arguments will need to be assessed, and the assessment will

[4] As I said, we are dealing with a plethora of epistemological problems concerning the revision of logic. It is not my goal in this contribution to discuss each problem in detail and try to reduce them to a unique form of epistemological problem. In this section, I just want to philosophically motivate the revision of logic and show that logic displays philosophical problems.

> need to be conducted in the context of some logic or other. But how do we choose this latter logic? We appear to be headed for an infinite regress. (p. 156)

Note that Agripa's famous trilemma, which challenges every single attempt to epistemologically ground our justifications and beliefs, may also be applied in this context of revision of logic. In Agripa's skeptical challenge, any chain of justification for our beliefs and theories will eventually face three problems. First, as presented in the quote above, once we get into the inquiry about the nature of logical principles and how to justify them, an infinite regress chain threatens us. We could avoid it by accepting a self-evident axiom or similar at the bottom of the hierarchy of justification. However, this strategy is viewed as arbitrary or dogmatic by a skeptic. Another way-out is to propose that beliefs and justifications should be grounded reciprocally, that is, they should be based on each other, as on a coherentist view. But if we use a net of evidence that should mutually ground each other, skeptics will point out that this might be viciously circular. Such arbitrariness in the stipulation of a first axiom, circularity or infinite regress challenges all foundational theories of knowledge. As Comesana and Klein put, "It is interesting to note that Agrippa's trilemma is perfectly general; in particular, it applies to philosophical positions as well as to ordinary propositions" (Comesana & Klein, 2004). In our view here, Agrippa's challenges also appear in the very discussion on the justification of our logical principles.

It is worth mentioning that the special philosophical predicates traditionally attributed to logic, 'necessity', 'universality', 'a priority' and 'eternity' in particular, also motivate a foundationalist approach to logic. According to epistemic foundationalism, the basic axioms, a privileged set of beliefs, ensure the epistemic security of all the steps taken in the system.

In what follows, we will develop a neopragmatist interpretation of this trilemma based on the intuition of Peregrin and Svoboda (2017), as they defend that:

> Building logical theories is thus best seen as a purpose-oriented enterprise and logicians should be viewed neither as visionaries having divine insight into 'the realm of the thought' nor as despots imposing laws on the proper use of language for those who want to be considered rational. They can instead be compared to democratic legislators who aspire to set some explicit rules of behavior but try to establish them so that they cohere with the current functioning of the society and with a wide social consensus regarding the worth of following them. (p. 5)

Like Peregrin and Svoboda, we hold that the process of successive modifications and revisions in a logical system is adequately expressed[5] in the notion of reflective equilibrium. The rules of logical systems should be seen as emerging from back-and-forth inquiry between facts about our already inferentially articulated discursive practices and provisional formulations of principles to, in a posteriori way, organize, limit and revise them. It is important to note that, in our neopragmatist view of revision of general principles, logic is primarily connected with our public and regulated linguistic activities and only marginally with formal rules and any rules of thought.

[5] This is made clear in the paper by the examples of revision that I am interested in here as in the intuitionist and paraconsistent traditions.

3 Revision of Logic and Normativity

We believe that two views on logic and its revision feature prominently in the current debate on the philosophical foundations of logic. The more hegemonic of the two can be taken as a naturalized approach that takes logic to be analogous to natural sciences, such as Physics. By highlighting representational and descriptive vocabulary, it provides a philosophically informed understanding of what logical theories do and the role that logical vocabulary plays.

Quine (1951, 1970) can be seen as the historical reference of this naturalized tradition. In his view, which has been very influential, even the laws of logic could be revised to reconcile one's system of belief with recalcitrant empirical data. This is the basis for what is now known in the literature as non-exceptionalism. Logic should not be taken as an a priori discipline and need not be regarded as normative since, as more general epistemological principles, not logical ones, could play the role of criteria to guide and correct inferential practices (Harman, 1984, 1986). Since we have representations and descriptions modeling different domains of reality, evidence and empirical adequacy are central to this hegemonic view. Thus, common strategies and methods to revise scientific theories, such as verification, regularity and relevance of evidence, falsifiability and induction, are decisive for the revision of logic (Williamson, 2017; Russell, 2012, 2014; Carnielli and Rodrigues, 2017). In addition, forms of abductivism and fallibilism are considered likewise (Priest, 2014; Bueno & Colyvan, 2004; Hjortland & Martin, 2017).

However, we can ask whether we should really engage with this naturalized view of logic to make its revision tenable. There is a second view, the normative one, which also highlights the possibility of the revision logic. This view does not, however, focus on central representationalist features. The normative approach to logic draws the analogy of logic with other normative theories and disciplines, the law, ethics and politics, for instance. This normative view of logic, although not hegemonic in contemporary research, displays a respectable history in philosophy. Frege (1879), for instance, seminally relates the nature of logic to the discussion of morals and freedom, when he states that:

> Logic has a closer affinity with ethics [than with psychology]... Here, too, we can talk of justification, and here, too, this is not simply a matter of relating what actually took place or of showing that things had to happen as they did and not in any other way. (Posthumous Writings, p. 4).

Other influential authors also seem to have defended approaches that assimilate logic to other normative disciplines rather than the natural sciences. Peirce is influential in defending that "logic is the ethics of thinking" (Kyriushchenko, 2011). Wittgenstein approaches the "sollen" of logic with the deontological "sollen" of ethics (1930, WWK p. 128, 131 and 175). Brandom (1994, 2001, 2008) holds that logical vocabulary expresses rules that articulate through a special "ought" our rational activities in the space of reasons. And Field (2009) defends the centrality of the normativity of logic and not representationalism to understand its nature.

In these normative approaches to the nature of logic, the central question of logic is not different from the central question of normative disciplines, namely: What should we do? What procedure should we adopt? What decisions need to be made? How should we guide our behavior in the world? Like the rules of normative disciplines, the rules of logic, before representing or describing domains of reality, are used by us to guide, correct, refine, revise what we do in the world.[6]

The analogy with ethical discussion is central to Prawitz in his *The Truth of Moral Propositions and Logic* (2007). In this work, he investigates the objectivity of our values and the possibility of submitting them to objective assessment. Prawitz wants to challenge the relativism and nihilism inherent in the discussion of moral values based on the need to have public criteria to evaluate moral judgments and draws the analogy between this investigation of the objectivity of moral values and the discussion on the choice of basic rules for logical systems. Is it possible to find a non-relativistic and objective procedure to justify the most basic principles of logic and allow for its revision?

In such a normative approach to logic, the method for revision of logic would not entail the fallibilism or abductivism typical of the approach to revise theories in natural sciences. The possibility of the revision of logic would, as in politics, for instance, be the method of reflective equilibrium, as defended by Prawitz (2007), Resnik (1997, 2004), Peregrin and Svoboda (2017).

Resnik (2004), for instance, mentions the importance of reflective equilibrium and traces the analogy of logic to normative disciplines as follows: "[Logicians] might follow the method of other normative theorists: they will seek a reflective equilibrium between their intuitions with regard to specific inferences and putative logical truths and their ongoing attempts to systematize the same using logical theories." (pp. 191–192) Developing Resnik's point, we may hold that when logic is taken as a normative discipline, we can supplement an anti-foundationalist approach to its epistemology with a normative proposal for the revision of logic grounded on reflective equilibrium. On this approach, it is worth noting, logic is a human and linguistic enterprise. Formal logic, reasoning using general principles, is regarded as based on our current and contingent discursive and inferential practices, more specifically local and everyday reasoning. The problem of justifying the most basic principles of logic is related neither to formal logic as it is usually understood nor to the application of logic to the natural sciences, but to our common everyday reasoning.

Consequently, the primary place of normativity and meaning is our daily lives, our regular inferential practices, as understood by Brandom (2008) and Peregrin and Svoboda (2017). The social dimension that constitutes logic is highlighted as well as the need to absorb rules by observation and instruction, the centrality of mutual correction in communal regulatory practices and the importance of training or initiation in collaborative practices through transmission and immersion in human communities (Dutilh-Novaes, 2016). As Resnik (2004) puts it: "Inference making is

[6] Here the analogy with normative disciplines comes from the recognition of the normativity of logic based on an inferentialist theory of meaning. In this view, normativity of logic is derivative of the normativity of our mastering of current concepts.

a social enterprise stabilized by the acquiescence, approval and disapproval of our fellows." (p. 188).

In this connection, it is worth examining how the notion of reflective equilibrium, informed by some neopragmatist tenets as expressivism and inferentialism, can be used to develop the possibility of revising logic, the dynamic and a posteriori aspect of such revision and the proper use of a normative bidirectionality in correcting general principles and local reasoning.

4 Reflective Equilibrium and Anti-Foundationalism

Before addressing the role of reflective equilibrium in the philosophy of logic, it is important to qualify our neopragmatist view of revising logic. Basically, our view concerns the socio-linguistic rules constitutive of our reasoning and seeks to reconstruct them systematically and explicitly as a formal system (Brandom, 1994, 2001; Peregrin, 2014). In other words, we look to the roots of logic in our linguistic interactions, especially to our mastering of concepts whose meaning is articulated by material inferences. We take it to be the work of logicians to explain and make explicit the constitutive rules and inferential transitions implicit in our linguistic practices and to simulate and systematize—with the help of mathematical methods—selected parts of real languages. Although logic contains much mathematics, it should start and end with pedestrian issues related to what humans do in daily life, that is, what native speakers of a natural language say and how they strive to perform their tasks and achieve their goals.

However, it is worth mentioning that this neopragmatist and normative view does not result in a commitment to the idea that logic must be empirical and count as unexceptional in comparison to natural science. Logical principles are not empirical generalizations since they do not represent the regularities in the use of our natural languages, but rather rules that *express* the norms inherent in the use of concepts that we master in our current lives (Brandom, 2001). This expressivist position presupposes that natural languages are essentially constituted rule-governed practices. Mastering a natural language is not just a matter of producing meaningful assertions. It consists, first of all, of normative attitudes—of entitlement and prohibition, for instance—in relation to the assertions of others and our own. We participate in a space of reasons that supports the implicit rules governing the use of linguistic expressions and signs and are, in this sense, constitutive of their meanings and senses (Brandom, 1994). For a semantic inferentialist, the "proto-logical" sphere is already dense in inferential relations, a fact that is only explained by a holistic view of logic. Logicians, in an expressivist view, should therefore try to make the implicit inferential structure of the natural language explicit in a formal system by regimenting and systematizing them. This view can be reconciled with Goodman's original ideas about reflective equilibrium (1958/1983). He did not coin the term 'reflective equilibrium' but was the first author to fully introduce and describe the procedure. He was trying to justify inductive procedures and address the problem of induction after Hume, namely, defend the proposition that local judgments and general principles should be properly balanced, that is, come into coherence with one another. The justification process is complex, since, according to Goodman, it consists of attempts to make mutual adjustments between accepted rules and inferences. The comprehensively achieved agreement is the only justification necessary for both accepted rules and local inferences.

4.1 *Reflective Equilibrium: State and Method*

Reflective equilibrium is characteristic of the daily practice of rational agents in their search for coherent and justified practices and theories. In order to do so, a rational agent should, in ethics or in politics, for example, examine their moral or political judgments in search of consistency with their beliefs and the broader domain of moral issues until an acceptable coherence between both is achieved by mutual and dynamical adjustment. This means looking for ways in which our beliefs can mutually support each other, that is, by reconciling sets of more varied beliefs, which, in turn, requires the revision and refinement of them at all levels in which challenges to some beliefs arise. We are to test various parts of our belief system against other parts in a back-and-forth procedure of epistemic evaluation and justification. Once beliefs are coherent with one another, particular judgments with general principles, they have a great degree of acceptability and credibility.

Accordingly, reflective equilibrium is the state to which we arrive or aspire to arrive in the process of developing our theories. In other words, it is the state in which theoretical principles we conjecture agree with the data we collect and vice versa. In these cases, we say we have reached a state of equilibrium based on mutual reflection and adjustments between general principles and local judgments. We could say that only systems of belief which achieve such equilibrium are indeed theories (Norman, 1979). Theoretical constructions that do not reach equilibrium are at best flawed theories.

However, reflective equilibrium is also a methodological principle. Even though it can be understood as a certain state we have reached or we aspire to reach, it can be seen as a method, a procedure the aim of which is to bring theoretical considerations and instances or particular cases into agreement. Norman (2023) aptly defines the method as follows:

> The method of reflective equilibrium consists in working back and forth among our considered judgments (some say our 'intuitions') about particular instances or cases, the principles or rules that we believe govern them, and the theoretical considerations that we believe bear on accepting these considered judgments, principles, or rules, revising any of these elements wherever necessary in order to achieve an acceptable coherence among them. (p. 2)

Norman (1979) introduces the idea of broad reflective equilibrium. In the logical case, as in the ethical case, we are looking for broad equilibrium. This means that we do not only want agreement among practical inferences and normative principles, but also their agreement with a broader background of theories and beliefs. Norman (1979) thus describes the mechanism in ethics by which judgments can be revised due to conflict with general principles and which, in turn, can be revised due to conflict with more local judgments.

> The method of wide reflective equilibrium is an attempt to produce coherence in an ordered triple of sets of beliefs held by a particular person, namely, (a) a set of considered moral judgments, (b) a set of moral principles, and (c) a set of relevant background theories. We begin by collecting the person's initial moral judgments and filter them to include only those of which he is relatively confident, and which have been made under conditions conducive to avoiding errors of judgment. [...] We then propose alternative sets of moral principles that have varying degrees of 'fit' with the moral judgments. We do not simply settle for the best fit of principles with judgments, however, which would give us only a narrow equilibrium. Instead, we advance philosophical arguments intended to bring out the relative strengths and weaknesses of the alternative sets of principles (or competing moral conceptions). These arguments can be constructed as inferences from some set of relevant background theories (I use the term loosely). It assumes that some particular set of arguments wins, and that the moral agent is persuaded that some set of principles is more acceptable than others. [...] We can imagine the agent working back and forth, making adjustments to his considered judgments, his moral principles, and his background theories. In this way he arrives at an equilibrium point that consists of the ordered triple (a), (b), (c). (p. 258f.)

However, if we use reflective equilibrium in a Brandomian framework to revise logic, we should criticize Norman's individualistic point of view. Norman seems to permit that the revision, at least in ethics, can be done by a particular person. But the same should not be the case in dealing with logic. In logic, we are talking about specialists, logicians trained in and recognized by a tradition that responds to pressures and challenges displayed by their communities in order to systematize and regiment inferential patterns that, in turn, we already master when we master the concepts in our natural languages. Bringing judgments into agreement means working with particular instances and cases, principles and rules that govern these cases and also with general theoretical considerations to adjust judgments, principles and rules. Thus, a judgment is justified if it coheres with the beliefs that a community (and its authorities) holds about right action—or correct inferences. Consequently, the logician who reconciles a principle or judgment in a state of reflective equilibrium with other beliefs should be justified by her community in believing that principle or judgment.

In the case of logic and its epistemology, Resnik (1997) takes the use of reflective equilibrium in logic to be analogous to other normative disciplines. He writes: "Reflective equilibrium applies to systems consisting of beliefs, logical theories, and considered judgments of logic. Such a system is in such a state just in case it is coherent by the lights of its own logical theory."[7] (p. 160) Later, in 2004, Resnik returns to the idea and develops the point as follows:

[7] For the question "who really does the reflective equilibrium", I would answer specialists. In the case of logic, the revision by reflective equilibrium is done by logicians whose authorities are recognized by their communities as they inquire relevant theories and pre-theoretical information in relevant contexts of natural languages. As an inferentialist holding logic as normative, I see meaning and normativity of logical systems as derivative of background commitments implicit in the use of current concepts in natural languages.

> Logicians start with their own set of accepted and rejected arguments — in Rawls's terms, their considered judgements of the facts of logic, that is, considered judgements of logical truth, validity, consistency, implication, and equivalence. Then they try to build a logical theory whose verdicts accord with their initial considered judgements. It is unlikely that their early attempts will produce an exact fit between the theory and the 'data'. Furthermore, they may produce a logical theory that proclaims unforeseen and prima facie anomalous logical relationships. Sometimes they can respond to such anomalies with a simple modification of their formal system. Sometimes retranslating a prima facie anomalous argument will reconcile it with the theory. Sometimes, however, they may yield their initial logical intuitions to powerful or elegant systematic considerations. In short, 'theory' will lead them to reject the 'data'. Furthermore, this mix of theory and 'data' often includes their other beliefs and commitments, including philosophical ones. The process comes to at least a temporary end when the logicians reach a state of 'reflective equilibrium', that is, one where they take their theory to reject no putative fact of logic that they are determined to preserve and to countenance none that they are determined to reject. (p. 181)

Following this neopragmatist way of thinking, in logic, as well as in ethics or politics, we can retrieve the idea of a back-and-forth movement between instances and theoretical models.[8] In the case of logic, these movements of coming and going involve concrete arguments in the natural language and the more general schemes expressed or made public in formal systems. By this means, formulae become closely connected to their instances and vice versa. So, for example, in contrast to other approaches to logic, a logical law like *modus ponens* does not have to be regarded as something discovered—whether in a hidden feature of the real world, an element of a hidden world of universals or an underlying a priori feature of our thinking. It can instead be seen as something that the logician constructs when working back-and-forth, on the one hand, on considerations about characteristics of concrete languages; on the other hand, on our more general theoretical attempts to express them. So, if one asks: a logician does the revision? I would say "yes". Such in the cases mentioned above of Da Costa, Brouwer and Priest. In this process, logicians immunize some generalizations against refutation and render them basic to our theories.

Peregrin and Svoboda (2017) offer the following example of the procedure of reflective equilibrium in relation to philosophical discussions regarding the legitimacy of the classic principle of the excluded middle:

> Consider, for example, the law of the excluded middle, stating, in effect, that every statement is true or false. In natural language, we can certainly find lots of statements which do not quite obey it: statements with indexicals or in other ways substantially context-dependent, vague statements, statements of personal taste, perhaps statements about the future, etc. So, there may seem to be a reason to reject the law and to embrace a logic that does not accept it. But, in view of the fact that two-valued logic is so conveniently simple, we may also consider the possibility to 'explain away' all the inconvenient cases as non-statements, or as not what we understand by statements when articulating the law. This, of course, is viable if there is a way of characterizing the inconvenient cases independent of the law (for saying that the

[8] For this analogy, I draw lessons on a theoretical model from Prawitz's work. As Prawitz put it, we use logical principles to correct local applications as we may use political principles and ethical ones to correct local ones, such as equality or any deontological principle. I just want to understand a phenomenon of philosophical import as the one of revising logic. As an inferentialist and expressivist in a Brandomian tradition, I think the notion of community is decisive.

> law applies to such and only such statements that do not violate it would, certainly, render it vacuous), but for at least some of them this might be possible. (We might, for example, want to exclude future tense statements on contingent matters, statements including some enumerated indexicals, etc.). (p. 95)

4.2 Some Epistemological Consequences of Reflective Equilibrium in Revising Logic

Tracing the epistemological consequences resulting from the method of reflective equilibrium, we see that we are dealing with justification rather than truth. General principles do not need to represent general and abstract states of affairs, but must be in agreement with one another, be coherent or even be reconcilable with particular instances and vice versa. In fact, in this view, we do not have to consider logical laws as absolute and invariable truths. Prawitz (2007) states:

> The existence of an equilibrium does not entail any absolute truth. But here we still find some criteria to adopt or to reject moral and logical propositions, namely, criteria for considering them true or false. Ultimately, even in the case of natural sciences, we do not have anything much better than that at our disposal. However, this is sufficient to take our rational discussions a step further also in the domain of moral and logical propositions. (p. 140)

According to this view, reflective equilibrium is based on a constant examination that allows the refining, modification and revision of logical systems. They can be modified as new elements emerge until stability is achieved. Such an anti-foundationalist approach to the epistemology of logic turns back the need for fundamental beliefs, principles or axioms that play the role of an ultimate foundation for research or knowledge. There is no need to construct a special epistemic history for basic rules of logic. There is no privileged or directly justified group of beliefs. There is, however, a fixed set of starting points, one that does not constitute a special foundation, apart from its relevance in our daily inferential practices. It is also worth noting that the background of the natural language, although contingent and historical, is not itself independent of logic. Natural languages are inferentially constituted and therefore implicated in logic from the beginning.

Once the epistemology of logic is based on non-foundationalism, we must face the problem of circularity. Does not the method of reflective equilibrium presuppose itself some logic?

4.3 Circularity, Anti-Foundationalism and Neurath's Boat

As we previously saw, Bueno and Colyvan (2004) believe that one of the reasons why the philosophy of logic is a difficult undertaking is that it presupposes what it studies, namely, logic. According to them, this difficulty is especially acute when we are dealing with revision of logic. The fact that we need logic to revise itself seems problematic. After all, it should result in circularity. This is an important epistemological problem.

According to the neopragmatist method proposed, revising the logic by the reflective equilibrium method is possible since deductive inferences can (and should) be justified by their conformity with intuitively valid general logical rules and general logical rules can (and should) be justified by their conformity with intuitively valid deductive inferences. The skeptic may argue that this dynamic and organic process of revision, the aim of which is to establish coherence between particular inferences and general logical rules or principles, is a circular process.[9] Indeed, while logical theory plays a normative role and must evaluate and correct our deductive inferential practices, the practices may come to exert a normative pressure on the rules of logical theory to the extent that they need to be modified. However, to modify these general logical rules and move towards a state of reflective equilibrium, we need a basic logic that enables and guides such action. Here again we see the issue of circularity. Circularity appears because we are using language to revise language. Logic to revise logic. And this is not a problem for an inferentialist *a la* Brandom and Peregrin, as I will defend below.[10]

It is worth noting that other authors have already pointed to this epistemological problem. Shapiro (2000), for instance, states that for the process towards reflective

[9] I use "coherence" in terms of Brandomian material compatibility (1994, 2008).

[10] As I said before, it is not my aim to defend Reflective Equilibrium against other detractors in this paper. My contribution here is about showing how Reflective Equilibrium can be used as a method to revise logic in a horizon where the normativity of logic is due to the normativity of natural language, as in a usual inferentialist point of view. Dicher (2022) and Woods (2019) present counter arguments to the use of Reflective Equilibrium in conjunction with forms of anti-exceptionalism. I am not an anti-exceptionalist. I hold that the procedure of logic revision is different from that of natural sciences. I see the tension between anti-exceptionalism and reflective equilibrium in Woods' paper. But my take on Reflective Equilibrium is not an-exceptionalist one. I do not think that logic is contiguous to empirical sciences. I have no problems with normativity to restrict the problem of freedom as presented by Woods. Regulations are necessary and they come from the normative and expressivist role that logics display. I agree with Woods that a Reflective-Equilibrium-based approach should be piecemeal. And this is not a problem. Changes in logic are in a sense conservative. Global conversions are (very) rare. There is no revolution but reforms. An inferentialist response to the problems of the permissibility as presented by Woods is outside the scope of this text, but it can already be pointed out to the inferentialist and expressivist retreat to natural language and consolidated linguistic practices. Such approaches thereby require some kind of guidance to separate out good repairs from bad ones. The guidance demanded by Woods is the expressivity and meaning of our inferential practices. One example given by Woods already set the inferentialist way-out: "If it's Wednesday, it's not Thursday." This is exactly the kind of material inference that an inferentialist takes as fixing the meaning of concepts (here, of days and weeks) and guiding the adjustments we make using Reflective Equilibrium in logic to express the normative meaning of some concepts we understand when we know how to use them. Dicher (2022) is not a problem either to an inferentialist informed way of thinking logic, because, in inferentialism, every revision should be held as public and normatively guided by the proto-logic of our daily concepts which are already inferentially articulated. In Brandom's inferentialism, for instance, there are, against Dicher's view, some pre-theoretical claims that should be informed by this "proto-logic" of material inferences that articulate the meaning of our concepts in natural languages. These material inferences that articulate the meaning of our concepts might guide our decisions in revising logics using reflective equilibrium.

equilibrium to start, we would need to have a logic operating in the background, but this is precisely what we are hoping to find by engaging in such a process. Moreover, Peregrin and Svoboda (2017), authors who propose reflective equilibrium, say:

> One of the most crucial problems of understanding logic in terms of the reflective equilibrium is the fact that we seem to need (at least a modicum of) logic to work toward the equilibrium —- and if logic were only to be the outcome of a process aiming at reflective equilibrium, we seem to be trapped in a vicious circle. (p. 95)

Accordingly, we may ask: is circularity really a problem? Is circularity something we should be concerned about in the revision of logic? Should we avoid circularity at all costs? To answer this objection to the method and the revision of logic, it is important to note that natural language already contains something like logic even before we begin to formalize it. This 'proto-logic' of material inferences that constitute our current conceptual practices transcends the process of reflective equilibrium and is too complex and confusing to admit a full formal reconstruction and regimentation.[11] Our natural languages are already immersed and constituted by dense inferential relations that determine our use of concepts (Brandom, 1994, 2001). As Peregrin and Svoboda (2017) defend:

> On the other hand, the logic that would be implicitly contained in a language might be — to a certain extent and in some respects — roam, indeterminate and open-ended, and to make it more precise, decisive, and explicit is a project that is surely not simple even if it may be greatly rewarding. This process is not just a description of what we can find in language, it is a project involving regimentation, streamlining, and extrapolation — but in no case is it a creation *ex nihilo*. (p. 96)

Prawitz (2007), when proposing reflective equilibrium, also observes that circularity is an inevitable problem. But it is not something that threatens the method. In fact, the author even argues:

> It seems that this way of arguing does not bring much to us; that it is just a circular reasoning. But, as I've tried to demonstrate, we face an inevitable circular movement here as we deal with human activities. [...] We can observe that this is a circular movement in which we can drastically modify both particular values and inferences and ethical and logical principles. When a conflict between a particular value and ethical principles, or between a particular inference and a general logical rule arises, we have to introduce some modification. If the validity of particular values and inferences is strongly rooted in us, it is likely that we reject the general principle that we consider wrong. If, on the contrary, the general principle is proved and we already absorbed it in our understanding of moral and logic, it is likely that we start to take the particular values and inferences as a mistake. (pp. 139–140)

For the purpose of claiming that circularity would not be a problem for our normative approach to logic, Resnik (2004) points out that:

---

[11] Here it is important to see that proto-logic presented in natural language is, by conception, not different from the logic under revision. As an expressivist, I hold the logical vocabulary displayed in logical theories as making explicit the conceptual and inferential meaning and normativity of different regions of this proto-logic. Again, this is an inferentialist lesson. Concepts in natural language are already articulated by (material) inferences.

> The worry about revising logic is a special case of the worry about revising some methodological norms while working within the very system of norms to which they belong. In particular, if something is a methodological apriority, one wonders how experience could ever lead us to review the methodological rules underwriting it. [...] Although it is difficult to see how a methodology for empirical theory testing can proceed without recognizing some fixed points, some apriorities, this doesn't mean that in developing a methodology or in discerning the so-called facts of logic we must depend upon apriori elements that transcend any methodology. (p. 184)

Soon after, Resnik (2004) says: "Instead of maintaining that we acquire norms through a priori means, I hold that we find ourselves with a collection of culturally conditioned norms and values, which we may or may not modify in the light of experience, arguments, and changes in our condition." (p. 185)

This quote shows Resnik's pragmatist view that is compatible with our view. Inspired by Prawitz and Resnik's line of argumentation, we argue that circularity would be a problem only for a foundationalist approach—in which the need to stipulate a set of special basic beliefs to accommodate the ascription of predicates as necessity, universality and apriority to logic.[12]

However, our pragmatist approach to thinking about the revision of logic starts from an anti-foundational view. According to this approach, logical principles need not be seen as something eternal, immutable, necessary, a priori, infallible and epistemically grounded on solid ground. In fact, logical principles are fallible, mutable and susceptible to revision, because they are the result of a long and complicated process. They were gradually established as specific forms of behavioral patterns acquired by competent language speakers. Thus, the logical principles must be taken as rules with normative power that constitute and correct our practices in the public discursive sphere of individuals in dynamic exchanges in the community.

According to our anti-foundationalist perspective, although no logical principle is necessary, all being revisable, we must rely on some to revise others. We always need principles playing the fundamental role of composing the basic logic needed to revise a principle. It is also worth mentioning that this foundation is obviously not eternal, in fact, it just needs to be stable enough to allow for our linguistic interactions. Thus, it makes no sense to question the validity and desire to review all the logical principles at once. In effect, the review should always take place gradually and locally.

Interestingly, the idea that we need logic to revise logic, and that revision does not take place at once, but gradually, is better understood given Neurath's anti-foundational metaphor, presented in his work *Anti-Spengler* (Neurath, 1921). According to his metaphor, we are like sailors who need to repair their boat on the high sea. Because there is no land close by, they cannot leave the boat to rebuild it. This can only be done gradually. When removing, say, any piece of wood, another one must be placed right away. For this to occur, the rest of the boat must be used as a support. In this way, it is possible to repair the boat completely, albeit gradually.

---

[12] I use inferentialism and expressivism to my take on Reflective Equilibrium to avoid the problem of circular justification. This is done with the analogy to Neurath's boat. Natural language is our first Neurath's boat.

If we look at the task of the logician who reviews theoretical considerations as being similar to the task of the Neurath's sailor who reconstructs his boat on the high seas, we will see that there is no problem with circularity. In fact, there is no other way to proceed organically and adopt a circular dynamic in which the logician uses some logic they are developing to modify that very logic. Just as the sailor cannot leave the boat, so the logician cannot "get out" of logic; there is no "out" of logic. Any kind of argument presupposes the use of logical principles that constitute our basic logic. In addition, just as the sailor cannot remove all the boards at once, so the logician cannot revise the logical principles all at once, there being in this event no criteria that would make the revision possible. This anti-foundationalist idea contrasts with a frequent foundationalist image in which it is necessary to completely demolish the building, at once, in order to rebuild it from a special foundation. As already argued, there must be some stable principles that serve as the foundation. However, this foundation is not immutable and can, therefore, admit of drastic change.

5 Normative Bidirectionality and Upward Normative Pressure

We have seen that the reflective equilibrium in logic proposed by Prawitz (2007), and other authors can be developed by analogy between logic and other normative sciences. Prawitz points out, for instance:

> In moral philosophy and in logic, we put our actions in a broader context, by trying to better understand its purpose and to find general rules for the activity as a whole. Because of that it is just natural that our evaluations and our way of reasoning are influenced by the principles that we reach in ethics and in logic, although these very principles are, in turn, the result of some reflection on this same activity and acquire its validity by being tested through it. (Prawitz, 2007, p. 139)

As a result, an approximation between logic and normative sciences, the reflection on values and our inferences and the attempt to formulate general principles for them is the start of a process of successive change. Such principles stand in relation to our previous practices and square with our theoretical explanations for it. This kind of equilibrium is a sturdy basis to accept the revision of logic, both in the case of local inferences and the case of the theory formulated about it based on general principles.

Pereira (2006) suggests a two-way path as a constitutive characteristic of this kind of revision:

> Our logical system arises from reflections on certain aspects of our human practices, in our analyzed case, of our inferential practices. When we theoretically reflect about our practices, and, in particular, on our deductive practices, these practices may be (and normally are) affected by our reflections. The relation between our practices and our theories concerning these practices is a two-way path: sometimes we evaluate, confirm or revoke theories because of its agreement or disagreement with some practices, other times we revise aspects of our practices through theoretical considerations. Our practices are not sacred: determined actions or arguments, taken before as morally acceptable or correct, may be revised, according to some light shed over our theories about these very actions and arguments. (p. 97)

In fact, there is a dynamic and organic aspect related to the revision of logic which I call here normative bidirectionality because it deals with reciprocal directions of normative orientation, correction, and pressure to change practices and rules. General principles play a normative role in guiding, revising and correcting local reasoning, but these, in turn, can be used to encourage or motivate changes and revisions of general principles as well. We can develop this notion of a two-way path introduced by Pereira in two forms: in the form of normative bidirectionality constituted by, at least, two forms of pressures, namely, upward and downward normative pressures.

General principles constraint and regulate local inferences. This normative direction of correction is clear and unproblematic for those who deal with theories of logic and its applications to domains of knowledge and everyday reasoning. There is, here, downward normative pressure to revise deviant cases of local inferences. I contend that, while the downward normative pressure is ubiquitous when we, for instance, teach logic to undergrads (i.e. to learn how to use general principles to change local applications of inferences), the introduced notion of upward normative pressure motivates some of the most interesting cases of a posteriori revision or restriction of logical principles (that is to pressure changes of general principles by the gained relevance of some local applications of inferences).

In this view, it is also important to see that local inferences may also exert pressure to constrain general principles. Particular inferences can motivate and encourage revision of more general rules. In short, local inferences, if salient, robust and rooted enough in our common inferential practices, can exert upward normative pressure to change general principles. Upward normative pressure takes place when the pressure to change the theoretical principles conjectured comes from the particular examples that we collected from our data, in other words, when the pressure comes from the ground up. As an inferentialist, data to exert upward pressure might be collected from our natural language use; as when we master the use of the current concept, we should master the material inferences drawn from this use. And as an expressivist, we use logical vocabulary to make explicit those inferential commitments that we master. Logicians, in turn, aptly organize, regiment and systematize these inferential patterns in formal systems. The dynamical combination and coherence between downward and upward normative pressures constitute what I call normative bidirectionality, which animates my view of reflective equilibrium, both as a method and as a result of this method.

The analogy between logic and normative disciplines to render my neopragmatist view of reflective equilibrium possible shows that general principles can be used to guide and correct particular instances and cases. In addition, it shows that these can be used to motivate their revision too. Indeed, the method of reflective equilibrium applied to logic implies that many interesting cases of revision occur upwards,[13] that is, the normative pressure for change occur from particular cases that gain relevance for a community to the understanding of the need to revise general principles and not

[13] The interesting cases here are at least the cases that I begin the paper with, namely intuitionism and paraconsistency. I am not defending Resnik's view, but I use his take to motivate my neopragmatist view influenced by inferentialism and expressivism.

the reverse, downwards, starting from general principles to correct local activities, as for instance, showing people that the affirmation of consequence and negation of antecedent while using conditionals are formal fallacies. It is worth noting, as Prawitz says (2007), the focus on general and *a priori* principles to understand logic generates infinite regress as denounced by Carroll (1895). Primarily, concrete activities themselves and not general principles for inferential activity should be at the heart of the discussion of the revision of logic.

According to Prawitz (2007), the problem with revising logic is not having true general principles. Perhaps these are important to make tough decisions in our moral lives, but it is not clear how to recognize the validity of some concrete reasoning (p. 135–136). It is also worth noting that the upward normative pressure, which particular cases exert for the modification of general principles, does not result in the view that we have empirical cases inductively determining general cases. In fact, we are correcting (relevant) practices with other (relevant) practices. General practices may have more normative weight in some circumstances and local practices may gain prominence in comparison to more general frameworks of our activities.

Let us look at some examples to illustrate the analogy and what we call upward normative pressure. In civil law in many countries, polygamous marriages are prohibited, that is, there are general legal principles which exert downward regulatory pressure on our practices, prohibiting, correcting and guiding them. Whoever has a polygamous marriage can be legitimately sanctioned. However, note that if, eventually, there are important cases of behavioral changes in our society over a significant period, this may encourage *a posteriori* revision of the general legal rule to accept polygamous marriages. In this case, there would be upward normative pressures exerted locally in society to the revision of general rules. This is also possible in logic, as we defend.

Before that, let us consider another example of upward normative pressure. In politics, likewise, a consolidated progressive or retrograde agenda among influential sections of society may generate upward normative pressure, that is, from particular cases to general principles, to introduce more inclusive general laws that welcome and benefit minority and historically oppressed groups or less inclusive laws that burden oppressed groups even more. This can also be seen in ethics. The strengthening of movements for human rights, women's rights and animal liberation, for instance, may encourage the change or revision of general ethical principles considered unfair. And the opposite is also possible. This change would once again be the result of an upward normative pressure.

As in the abovementioned examples, the upward pressure is exercised by people held as relevant for the corresponding activity while their practice gains more and more traction, recognition and centrality for an overall concerned community whether in politics, law or in ethics. Analogously to such other normative disciplines, in logic, local discussions of relevant people whose practices are recognized specialists[14] such

[14] Here, I would like to say some words concerning the issue of collecting data. I am not holding that "collecting data" means "collecting through a survey of what people understand to be valid arguments". This methodology seems to be inadequate, as pointed out by an anonymous reviewer,

as, for instance, about meaning, may prompt the restriction of general principles of logic such as in the case of intuitionism, wherein the restriction of the Excluded Middle based on the examination of the assertability conditions of language is encouraged (Dummett, 1977, 1978).[15] We could also think of cases of vagueness, indexicality and future contingents that upwardly motivate the modification, restriction, reorientation or even the rejection of logical principles that were previously considered universal and necessary. And this pressure renders philosophical motivation for new non classical logics that restrict or revise classical principles.

Some paraconsistent logics, for example, may also be the result of this kind of upward normative pressure. Conflicting information cases have gained prominence in increasingly complex human information exchanges, so that it becomes important to restrain the universality of the classical Principle of Explosion in order not to trivialize the relation of logical consequence through contradictory information (Carnielli & Rodrigues, 2019a, b). In this sense, local, but important, cases of conflicting information in belief systems or databases generate upward normative pressure, that is, the need to correct or revise the universality of a classical logical principle, as the one that states a system with a contradiction is a trivial one.

6 Concluding Remarks

My contribution here focused on showing how Reflective Equilibrium can be used as a method to revise logic in a horizon where the normativity of logic is due to the normativity of natural language, as in a usual inferentialist point of view. We have highlighted, using inferentialist and expressivist insights, the normative character of logic, the dynamic aspect of the revision of logic, the normative bidirectionality and the upward and downward normative pressures in the use of reflective equilibrium in logic based on some features familiar to other normative disciplines. We proposed a neopragmatist and anti-foundationalist view to the epistemological problem concerning logic. It is, we argued, possible to avoid the false dilemma of choosing between a metaphysical or strictly arbitrary philosophical basis for the foundation

because many people, for instance, believe that some formal fallacies can be a valid argument. Even if the majority of the people interviewed in the survey replied that this is a valid inference, says my referee, we *should not* be inclined to change this logical principle. I agree with this remark. Not every inferential pattern used by any person who masters a natural language plays a relevant role for this sort of upward pressure. We should then sort out inferential patterns of specialists and also of naive people. In this paper, I just want to defend the need to acknowledge this upward normative pressure to understand revision of logical principles and not how this is currently done. In order to do so, we need more empirical investigation, I think. How a person becomes a recognized specialist, or its group gains more voice, relevance and traction in social exchanges among specialists is an interesting question, but it is beyond the scope of this paper.

[15] Here it is worth noting that there is no necessary upward revision because of a philosophical position, but there is *pressure* to revise exerted by a philosophical position. That is enough for my point.

of the basic rules of logic. Circularity is also not a threat to revise logics.[16] Our neo-pragmatist view of revising logic appeals to reflective equilibrium as a conceptual platform that provides a philosophically interesting alternative way of appreciating how logic can be revised. Reflective equilibrium does not assume the existence of a metaphysical basis, nor does it presuppose a plain conventionalist approach to logic. Additionally, we have also shown that it can be rich enough to express the process of revising various types of non-classical logics.[17]

There are grounds for the most basic rules of logic, but they are neither a priori nor absolute. It is as contingent and a posteriori as other human institutions. We are educated and introduced to a natural language already dense with inferential relations, relations that constitute our everyday concepts and form a crucial part of our so-called logical principles. Our justification system is important in our education, that is, our initiation into an inferential practice and it is always related to a predetermined discursive practice. It is no accident that rule-governed practices play a central role in any pragmatist approach to linguistic phenomena. Logical principles are expressions of our regular practices typically established over a long period of complicated cooperative activities. Norms unite us in our linguistic interactions. They are expressed in different formal systems and can be different as well as change. On this view, norms are heterogeneous. We are not merely dealing with prescriptions and maxims, but rather with the consolidation of standards from which we judge our actions and descriptions to be correct or not. This shows how to make logic closer to politics, because logic, among other reasons, is also susceptible to many different normative pressures as any other normative activities. What I call upward normative pressure adequately expresses the dynamical and a posteriori aspect of the revision of some logical principles.

---

[16] For one who might say that the issue is not simply revising logic, and it is required to show that revisions are rational, I would say that it depends on the notion of rationality one is using. I am using a Brandomian expressivist view of rationality. But to develop this view here is beyond the scope of this contribution.

[17] I do not see pragmatism as a methodology, as pointed out by a referee, but rather as a philosophical tradition. My contribution makes explicit my affiliation to this view. I am holding a pragmatist view of Reflective Equilibrium as rich enough to deal with many different heterogeneous philosophical motivations, as the one displayed by Brouwer e and others about Brazilian paraconsistency tradition. And my main topic here is on the epistemology of logic and how to combine Reflective Equilibrium with pragmatism. I use reflective equilibrium pragmatically to deal with two problems I presented related to a dilemma between conventionalism and metaphysics. I do not hold that the result is pragmatism because of Reflective Equilibrium. It is pragmatist because of the tradition from which I am reading reflective equilibrium. I see my contribution concerning philosophy (theory of meaning and expressivism) and not concerning prima facie methodology. Regarding the motivation to use pragmatism, I believe it is linked to the inferentialism and expressivism highlighted at the beginning of the text. It is an important contemporary tradition that can avoid conventionalism a la Carnap and metaphysics. I believe the conclusion about normative pressures is a good way to show why I am a pragmatist. I am not a pragmatist because I use reflective equilibrium. Rather my use of Reflective Equilibrium makes explicit my pragmatist assumptions. Reflective equilibrium can be made compatible with Brandom's inferentialism and expressivism.

Acknowledgements This research was funded by a grant of the National Council for Scientific and Technological Development (CNPq/Brazil). Marcos Silva is Productivity Researcher Level 2 of CNPq.

References

Brandom, R. (2001). *Articulating reasons: An introduction to inferentialism*. Cambridge: Harvard University Press.

Brandom, R. (1994). *Making it explicit: Reasoning, representing and discursive commitment*. Cambridge: Harvard University Press.

Brandom, R. (2008). *Between saying and doing: Towards an analytic pragmatism*. Oxford: Oxford University Press.

Brouwer, L. E. (1907). *Over de Grondslagen der Wiskunde*. Ph.D. Thesis, Universiteit van Amsterdam. English translation in Brouwer 1975: pp. 11–101. Onbetrouwbaarheid der Logische Principes. Tijdschrift voor Wijsbegeerte, 2: pp. 152–158. English translation in Van Atten and Sundholm 2017.

Bueno, O., & Colyvan, M. (2004). Logical non-apriorism and the law of non-contradiction. In *The law of non-contradiction: New philosophical essays* (pp. 156–175).

Carnielli, W., & Rodrigues, A. (2015). Towards a philosophical understanding of the logics of formal inconsistency. *Manuscrito, 38*(2), 155–184.

Carnielli, W., & Rodrigues, A. (2016). On the philosophy and mathematics of the logics of formal inconsistency. In J.-Y. Beziau, M. Chakraborty, & S. Dutta (Eds.), *New directions in paraconsistent logic* (pp. 57–88).

Carnielli, W., & Rodrigues, A. (2017). An epistemic approach to paraconsistency: A logic of evidence and truth. *Synthese, 196*, 3789–3813.

Carnielli, W., & Rodrigues, A. (2019). On epistemic and ontological interpretations of intuitionistic and paraconsistent paradigms. *Logic Journal of the IGPL*.

Carnielli, W., & Rodrigues, A. (2019). Inferential semantics, paraconsistency and preservation of evidence. In C. Baskent, & T. Ferguson (Eds.), *Graham Priest on dialetheism and paraconsistency*. Springer, Berlin.

Carroll, L. (1895). What the tortoise said to Achilles. *Mind, 4*(14), 278–280.

Comesana, J., & Klein, P. S. (2004). In E. N. Zalta & U. Nodelmann (Eds.), *The stanford encyclopedia of philosophy* (Spring 2024 Edition). https://plato.stanford.edu/archives/spr2024/entries/skepticism/

da Costa, N. (1958). Nota sobre o conceito de contradição. In *Proceedings of the Paraná Society of Mathematics*, vol. 1, pp. 6–8. [in Portuguese].

da Costa, N. (1959). Observações sobre o conceito de existência na Matemática. In *Proceedings of the Paraná Society of Mathematics*, vol. 2 (pp. 16–19). [in Portuguese].

da Costa, N. (1974). On the theory of inconsistent formal systems. *Notre Dame Journal of Formal Logic, 15*, 497–510.

Daniels, N. (1979). Wide reflective equilibrium and theory acceptance in ethics. *The Journal of Philosophy, 76*(5), 256–282.

Daniels, N. (2023). Reflective equilibrium. In E. N. Zalta & U. Nodelmann (Eds.), *The stanford encyclopedia of philosophy* (Fall 2023 Edition). https://plato.stanford.edu/archives/fall2003/entries/reflectiveequilibrium/

Dicher, B. (2022). Reflective equilibrium on the fringe: the tragic threefold story of a failed methodology for logical theorising. *Dialectica, 999*(1).

Dutilh-Novaes, C. (2015). A dialogical, multi-agent account of the normativity of logic. *Dialectica, 69*, 587–609.

Dutilh-Novaes, C. (2016). Reduction ad absurdum from a dialogical perspective. *Philosophical Studies, 173*(10), 2605–2628.
Dutilh-Novaes, C. (2020). *The dialogical roots of deduction*. Cambridge: Cambridge University Press.
Dummett, M. (1977). *Elements of intuitionism*. Oxford: Oxford University Press.
Dummett, M. (1978). *Truth and other enigmas*. Cambridge: Harvard University Press.
Field, H. (2009). What is the normative role of logic? *Aristotelian Society Supplementary, 83*, 251–268.
Frege, G. (1879). Halle: Nebert; English translation "Begriffsschrift". In J. van Heijenoort (Ed.), *From Frege to Gödel: A source book from mathematical logic* (pp. 1–82). Cambridge: Harvard University Press.
Goodman, N. (1955). *Fact, fiction and forecast*. Cambridge: Harvard University Press.
Hacker, P. (1972). *Insight and illusion*. Oxford: Clarendon Press.
Harman, G. (1984). Logic and reasoning. *Synthese, 60*(1), 107–127.
Harman, G. (1986). *Change in view: Principles of reasoning*. Cambridge: MIT Press.
Heyting, A. (1928). *Essay on the formalization of intuitionistic logic*. Expanded and revised version published as Heyting 1930, Heyting 1930a, Heyting 1930b.
Heyting, A. (1930). Die formalen regeln der Intuitionistischen Logik I. *Sitzungsberichte der Preussischen Akademie der Wissenschaften* (pp. 42–56). English translation in Mancosu, 1998 (pp. 311–327).
Heyting, A. (1930a). Die formalen regeln der Intuitionistischen Logik II. *Sitzungsberichte der Preussischen Akademie der Wissenschaften*(pp. 57–71).
Heyting, A. (1930b). Die formalen regeln der Intuitionistischen Logik III. *Sitzungsberichte der Preussischen Akademie der Wissenschaften* (pp. 158–169).
Hjortland, O. (2017). Anti-exceptionalism about logic. *Philosophical Studies, 174*(3), 631–658.
Kyriushchenko, V. (2011). Logic, ethics and aesthetics. *European Journal of Pragmatism and American Philosophy* [on-line]: III-2. http://journals.openedition.org/ejpap/849
Marcos, J. (2004).*Logics of formal inconsistency*. Thesis (doctorate) - State University of Campinas.
Marcos, J. (2010). Wittgenstein & paraconsistency. *Principia, 14*(1), 135–173.
Neurath, O. (1921). Anti-Spengler. In R. S. Cohen & M. Neurath (Eds.), *Empiricism and sociology*. Dordrecht: Reidel.
Peregrin, J. (2014). *Inferentialism: Why rules matter*. Basingstoke: Palgrave.
Peregrin, J., & Svoboda, V. (2017). *Reflective equilibrium and the principles of logical analysis: Understanding the laws of logic*. New York: Routledge.
Pereira, L.C. (2006). Brief considerations on nihilism and revisionism in logic. *What makes us think,20*, 91–99.
Prawitz, D. (2007). On the truth of moral propositions and propositional logic. Trad. Luiz Carlos Pereira. *Analytica Magazine, 11*(1), 211–233.
Priest, G. (2014). Revising logic. In P. Rush (Ed.), *The metaphysics of logic* (pp. 211–233). Cambridge: Cambridge University Press.
Priest, G., Tanaka, K., & Weber, Z. (2015). Paraconsistent logic. In E. N. Zalta & U. Nodelmann (Eds.), *The Stanford encyclopedia of philosophy* (Spring 2015 Edition). http://plate.stanford.edu/archives/spr2015/entries/logicparaconsistent/
Quine, W. (1951). Two dogmas of empiricism. *The Philosophical Review, 60*, 20–43.
Quine, W. (1970). *Philosophy of logic*. Cambridge: Harvard University Press.
Rawls, J. (1999) *A theory of justice* (2nd ed.). Cambridge: Harvard University Press.
Resnik, M. (1997). *Mathematics as a science of patterns*. Oxford: Clarendon Press.
Resnik, M. (2004). Revising logic. In G. Priest, J. C. Beall, & B. Armor-Garb (Eds.), *The law of non-contradiction* (pp. 178–194). Oxford: Clarendon Press.
Russell, G. (2014). Metaphysical analyticity and the epistemology of logic. *Philosophical Studies, 171*(1), 161–175.
Russell, G. (2012). The justification of the basic laws of logic. *Journal of Philosophical Logic, 44*(6), 793–803.

Shapiro, S. (2000). In P. Boghossian, & C. Peacocke *New essays on the a priori* (pp. 333–366). Oxford: Clarendon Press.
Williamson, T. (2014). Logic, metalogic and neutrality. *Erkenntnis, 79*(2), 211–231.
Williamson, T. (2017). In B. Armour-Garb (Ed.), *Reflections on the liar* (pp. 325–346). Oxford: Oxford University Press.
Warren, J. (2020). *Shadows of syntax: Revitalizing logical and mathematical conventionalism*. New York: Oxford University Press.
Wittgenstein, L. (1967). In B. McGuinness (Ed.), *Wittgenstein und der wiener kreis*. Gespräche, Frankfurt: Surkhamp.
Woods, J. (2019). Against reflective equilibrium for logical theorizing. *Australasian Journal of Logic, 16*(7), 319.

Logic Is Not Science

Ulf Hlobil

Abstract I argue that logic is unlike science in its methodology, thus rejecting anti-exceptionalism about logic. Logic has a mathematical and a philosophical part. In its mathematical part, the methodology of logic is like that of mathematics, and no need to choose between theories arises in that part. In its philosophical part, the methodology of logic is like that of philosophy. Philosophy and mathematics are both unlike the empirical sciences in their methodology. So logic is unlike the empirical sciences in its methodology. I end by looking at disagreements about the liar paradox as an example.

1 Introduction

> [L]ogic is, at root, a philosophical enterprise. Since at least the beginning of the twentieth century, however, logic has become a branch of mathematics as well as a branch of philosophy. (Shapiro, 2005, 651)

I agree with what Shapiro says here; and, in this chapter, I want to bring out some implications of this view for debates about the methodology of logic and theory choice in logic, namely by arguing for the following view: Logic is unlike the empirical sciences in its methodology. Logic has a mathematical part that is like mathematics in its methodology, and it has a philosophical part that is like philosophy in its methodology. In the mathematical part of logic, no need to choose between different logical theories arises. In the philosophical part of logic, there are disagreements between logical theories, so that the question of rational theory choice arises. However, the philosophical part of logic shares characteristic methodological features with philosophy. Since the methodologies of mathematics and philosophy are exceptional, it follows that the methodology of logic is exceptional. So, anti-exceptionalism about logic is false.

U. Hlobil (✉)
Concordia University, Montreal, Canada
e-mail: ulf.hlobil@concordia.ca

S. Molick (ed.), *Demarcating Logic and Science*, Studies in Applied Philosophy, Epistemology and Rational Ethics 80, https://doi.org/10.1007/978-3-032-25674-4_7

The chapter is organized as follows: In Sect. 2, I argue that anti-exceptionalism is unclear as it stands and should be clarified in such a way that it implies that the methodology of logic is unlike that of mathematics or philosophy. In Sect. 3, I argue that the mathematical part of logic is like mathematics in its methodology. In Sect. 4, I begin my argument that the philosophical part of logic is like philosophy. And I then support this claim further by looking at a disagreement about the liar paradox in Sect. 5. Section 6 concludes.

2 What Is Anti-exceptionalism?

I am writing in opposition to the current trend of anti-exceptionalism in the philosophy of logic. Recently many philosophers who otherwise share very few commitments have endorsed anti-exceptionalism about logic, and it is perhaps the currently most popular view about the nature and role of logic (Ferrari et al., 2023; Woods, 2023; Tajer, 2022; Martin and Hjortland, 2022; Carlson, 2022; Arenhart, 2022; Molick, 2021; Peregrin and Svoboda, 2021; Hjortland, 2019; Russell, 2019; Read, 2019; Finn, 2019; Costa and Arenhart, 2018; Wyatt and Payette, 2018; Williamson, 2017; Priest, 2014; Quine, 1964).[1] In this section, I first argue that the claims of anti-exceptionalism are not clear enough to be assessed as they stand. I then suggest that, at a minimum, anti-exceptionalists should hold that the methodology of logic is unlike that of mathematics and philosophy.

Hjortland (2017, 631) formulates anti-exceptionalism thus:

> Logic isn't special. Its theories are continuous with science; its method continuous with scientific method. Logic isn't *a priori*, nor are its truths analytic truths. Logical theories are revisable, and if they are revised, they are revised on the same grounds as scientific theories.

According to this, anti-exceptionalists should hold that logic is not *a priori* and probably revisable in light of discoveries about nature. However, some philosophers who self-identify as anti-exceptionalists hold that logic might be *a priori* (Russell, 2019), and others are non-naturalists (McSweeney, 2021). Hjortland also suggests that exceptionalists hold that logical theory is unrevisable and analytic. However, while Kant (KrV, B viii) notoriously held that logic has made no progress since Aristotle, very few philosophers of logic or logicians agree (including exceptionalists). As Kneale and Kneale (1962, 355) say, Kant "was apparently unaware [...] that the doctrine which he regarded as the complete and perfect discovery of Aristotle was in fact a peculiarly confused version of the traditional mixture of Aristotelian and Stoic elements." Similarly, the notion of analyticity is notoriously difficult to clarify (Quine, 1951), and hardly any philosopher or logician today (including exceptionalists) would put a lot of weight on this notion without sharpening it in specific ways (see Russell, 2008). So it seems that some of the ideas in Hjortland's formulation are

[1] Recent criticism is offered by Kripke (2023), Andersen (2023), Sagi (2021), Rossberg and Shapiro (2021), dos Santos (2021), Dicher (2020), and Woods (2019a, 2019b). My worries overlap with some of these other criticisms but are not identical to any of them.

not universally accepted by anti-exceptionalists, and others are too unclear or controversial to function as core commitments of an exceptionalist opponent. Once we ignore these problematic parts, it seems that the core claims of anti-exceptionalism are that logical theories and methods are continuous with—or relevantly similar to those of—science, and that logical theories are revised (or chosen) on the same grounds as scientific theories.

Unfortunately, it is not obvious what "continuous" or "relevantly similar" and "science" mean in this claim (Rossberg and Shapiro, 2021). I will take these issues in turn and start with "continuous" or "relevantly similar." Then I will turn to what anti-exceptionalists should mean by "science."

2.1 "Similar" in What Way?

Sometimes the idea that logic is "continuous" or "relevantly similar" to science is taken to assert a similarity in the nature of the studied entities. More often, however, the respect of similarity is taken to concern epistemology and methodology (Martin and Hjortland, 2022).[2] Understood as a methodological thesis, anti-exceptionalism is taken to imply that theory choice in logic happens in a way that is similar to how theory choice happens in the sciences.[3] Some think that this means that theory choice happens by abduction or inference to the best explanation (Priest, 2016, 2020). Others think that theories are chosen according to their power to make (confirmed) predictions (Martin and Hjortland, 2020). Yet others hold that theory choice in logic is like the choice of models in empirical sciences (Arenhart, 2022).

All of these views share the idea that logical theories are assessed in light of some data that the theory ought to relate to in an appropriate way, namely by explaining, or predicting, or modeling them. The term "datum" suggests that something is given to us (more or less) independently of our logical theorizing, so that we can compare and choose theories in light of what is given. Unfortunately, it is not easy to say what is "given" in such a way that we can compare and choose logical theories in light of it (Hlobil, 2020). Some think that what is given are intuitions about the goodness of arguments in natural language (Priest, 2016, 2020). Others think that what is given are very general facts about everything (Williamson, 2017). Yet others think that what is given are coherence norms governing discourse (2015, Ripley, 2017a). And some think that what is given are social facts about how mathematicians assess each other's proofs (Burgess, 1992). Moreover, as already intimated, while some think that what is given is *a posteriori* (Hjortland, 2017, 631) others think that some of the given may be *a priori* (Russell, 2019). So what is given to us in our logical theorizing

[2] Since I am focusing on epistemic and methodological anti-exceptionalism about logic, I do not discuss the relation between logical anti-realism or instrumentalism and logical anti-exceptionalism in any detail here (see Tahko, 2021; Commandeur, 2024).

[3] However, some philosophers who self-identify as anti-exceptionalists explicitly reject the continuity of methodology between the empirical sciences and logic (McSweeney, 2021).

and how it is given to us are controversial issues. Indeed, one might suspect that the idea that there is a "given" in logic that plays a role comparable to that played by experimental results and observations in the empirical sciences is a myth, which we may call the "myth of the logically given."[4]

Perhaps an opponent would want to dispel the myth of the logically given by pointing out that it is almost universally accepted that arguments that have the form of *modus ponens*, universal instantiation, Barbara and the like are logically valid, while arguments that have the form of affirming the consequent are not logically valid on account of this form. Can't we use what we thus agree on as "the given" in logical theorizing? While it may be useful to have some agreement on the extension of logical validity, the problem is that there is no agreement on what it means for an argument to be logically valid.[5] Does the validity of, say, *modus ponens* mean that it is necessarily truth preserving (if so, what does "necessarily" mean here)? Or that it codifies a good kind of reasoning? Or that any body of information that includes the premises also includes the conclusion? Or that it is incoherent to assert the premises and deny the conclusion? Or that mathematicians accept it as a step in a mathematical proof? Or something yet different? When logical theorizing gets hard, these differences start to matter. When we want to compare and choose between logical theories, we must usually consider cases that are not clear cut and uncontroversial and for which the question what it means for an argument to be logically valid makes a difference (Hlobil, 2020; Dicher, 2020). Once we face issues like the semantic paradoxes, for instance, we must give up something widely accepted; and which principles regarding validity it makes sense to give up may (and indeed often does) depend on what it means to accept it as valid in the first place. I will come back to this point below.

Partly in light of how difficult it is to articulate what "relevantly similar" means, some anti-exceptionalists have started to use "anti-exceptionalism" for any position that denies that logic is general, formal, foundational, *a priori*, analytic, and necessary (Martin and Hjortland, 2022). According to this understanding, anti-exceptionalists need not agree on anything except on a gerrymandered disjunction of claims, namely that logic is either not general or not formal or not foundational, etc. It seems to me that the disjuncts of this claim deserve to be discussed separately, so that the label "anti-exceptionalism" becomes unhelpful when it is understood in this way.

Instead of saying what it means to be "relevantly similar" to science, a more promising approach might be to say what it means *not* to be "relevantly similar" to science. Martin and Hjortland (2024) suggest that the two alternatives to an anti-exceptionalist ("science like") methodology for logic are (a) that rational theory choice in logic happens by unmediated rational insights (logical rationalism) or (b) that it happens by appreciating the analyticity of logical principles in virtue of grasping the meanings of logical expressions (semanticism). However, if these are the only

[4] Arenhart (2022) holds that the data are "co-constructed" with the target phenomenon that one aims to model by a logical theory. It is, hence, not clear to me whether Arenhart accepts the myth of the logically given.

[5] Some prominent philosophical logicians deny that logical theories are theories of logical validity or consequence (Williamson, 2024). I ignore this extra problem for my opponent here.

genuine alternatives to an anti-exceptionalist methodology of any discipline, then any arguments that speak against rationalism and semanticism in general also speak for anti-exceptionalism in general, with respect to any discipline. Against rationalism, Martin and Hjortland (2024) say that it is unclear whether we have any capacity for rational insights and whether it can yield knowledge. And against semanticism, they point to general doubts about analyticity and links between understanding and assent. These arguments generalize to other fields, and in particular to philosophy and mathematics (which will become relevant below). So, if the arguments that Martin and Hjortland present are cogent, then their conclusion should not be anti-exceptionalism about logic in particular but rather anti-exceptionalism about philosophy, mathematics, and probably all disciplines.[6] Therefore, Martin and Hjortland's arguments, if cogent, either support a global version of anti-exceptionalism or there is an alternative methodology besides rationalism, semanticism, and anti-exceptionalism.

Global anti-exceptionalism is an interesting claim only if there are plausible methodologies that some discipline might follow but none actually follows. Otherwise, global anti-exceptionalism degenerates into the uninteresting claim that all disciplines are alike in some unspecified and vague sense. Unfortunately, the versions of rationalism and semanticism that Martin and Hjortland sketch are not methodologies that are plausible alternatives to the alleged science-like method, which would give us a clear idea of what anti-exceptionalism is denying. So Martin and Hjortland should make sure that their anti-exceptionalism does not degenerate into global anti-exceptionalism. They owe us an account of what "science" is, such that there may plausibly be disciplines that are unlike science in their methodology. This brings us to the question what anti-exceptionalists should mean by "science."

2.2 *"Science" in What Sense?*

Anti-exceptionalists owe us an account of what they mean by "science" or "the methodology of the sciences" that makes their claims worth considering. Since the

[6] Some philosophers hold that philosophy is or ought to be similar in its methodology to "science" (where this is usually taken to mean something like "natural science"). For example, Williamson (2013a) explicitly endorses exceptionalism about philosophy (and not just about logic). Williamson therefore owes us an account of what it would be to be exceptional. I think what he says on this point is not enough to make anti-exceptionalism interesting. While his *The Philosophy of Philosophy* (Williamson, 2007) includes detailed criticism of the linguistic turn and ordinary language philosophy, his positive proposal about the methodology of philosophy remain so thin as to amount to little more than the suggestion that we try to consider issues carefully and rationally in philosophy, which I do not think is an interesting claim about the methodology of philosophy.

Other philosophers also hold a more wide-ranging anti-exceptionalism. Ladyman and Ross (2007) hold that metaphysics should not go beyond unifying different parts of (natural) science. A little bit more liberal is Emery's (2023) view that metaphysics ought to be informed by science and that its method is continuous with the methods used in science. I set such views aside here because my target is anti-exceptionalism about logic, independently from more global kinds of anti-exceptionalism.

quest for a demarcation criterion for science has failed, it is unclear whether the sciences share any particular methodology (Laudan, 1983). Moreover, Tajer (2022) has shown that we find diverse methodologies in logic. So instead of focusing on methodology, perhaps we can make progress on what anti-exceptionalism is by considering which disciplines should count as sciences.

There is a very broad understanding of "science" on which any discipline that offers theories of some subject matter is a science; and on this understanding science contrasts with the other four Aristotelian intellectual virtues: skill, prudence, rational insight, and wisdom. If "science" is accordingly taken to mean something like the Latin *scientia*, the German *Wissenschaft*, or the Greek *episteme*,[7] then one of the most extensively discussed questions in the philosophy of logic over the past two millennia is whether logic is a science in this broad sense. This traditional disagreement centered on the question of whether logic is in the business of offering theories of some subject matter, or whether it is rather a skill or art, in the business of regulating or improving some activity (as it was universally agreed that logic cannot be part of prudence, rational insight, or wisdom). However, this is not the question that anti-exceptionalists have in mind. They do not want to contribute to this traditional debate. Rather, they seem to understand "sciences" as disciplines that share a particular method of rationally choosing theories. So how must theories be chosen in a discipline for the discipline to count as a science in this sense? We are thus led back to the issue of methodology.

Martin (2020) has suggested that the relevant anti-exceptionalist method consists of abduction or inference to the best explanation (though in other writing he claims that the method is that of selecting theories based on the predictions they make (Martin and Hjortland, 2020)). In particular, Martin looks at the work of Priest (2006) as a case study and argues that Priest's methodology is abductivist because it relies on the following assumptions:

(i) Good theories solve open puzzles in their field (for example, the semantic paradoxes in logic).
(ii) Such solutions are better if they can be motivated on independent grounds.
(iii) Independent motivations include the ability to solve other puzzles and account for important further commitments.
(iv) Commitments that should be taken into account when assessing the quality of a solution include the (apparent) meaningfulness of sentences, the constitutive properties of concepts, and the soundness of results and practices in established disciplines.

[7] As an aside, it is worth pointing out that most anglophone philosophers today use "science" in a way that is in sharp contrast with how the German speaking tradition of the Vienna circle, Popper, or Feyerabend used the term *Wissenschaft*, which is much closer to *scientia* and *episteme*. The concept used by anglophone philosophers seems to be a peculiar result of combining the Renaissance distinction between letters and sciences and the 19th century German distinction between *Geisteswissenschaft* and *Naturwissenschaft*. I doubt that the currently common English terminology is philosophically helpful; but I won't engage the issue here.

(v) Such commitments can make a difference to whether one ought to accept the solution (and they are, in this sense, evidence for or against such solutions).

So should we perhaps understand "science" as any discipline whose methodology relies on these five assumptions? This suggestion degenerates into global anti-exceptionalism. The problem with this idea is that assumptions (i)–(v) are made in virtually all intellectual (theory seeking) human endeavors. If making assumptions (i)–(v) suffices to employ an anti-exceptionalist methodology, then the methodology of logic is almost certainly anti-exceptionalist—and so is pretty much every methodology for an intellectual discipline anyone ever thought about.

So far, we have not found any illuminating option for what anti-exceptionalists could mean by "science" or "science-like." The problem is that the methodologies that anti-exceptionalists have specified are too broad and too vague to render their core claim interesting. I want to suggest that in order to be an interesting claim, anti-exceptionalists should exclude some disciplines from what they mean by "science," so that their core claim implies that the methodology of logic is not like that of those disciplines. Moreover, the disciplines that are not like science should be such that someone might reasonably think that their methodology is similar to the methodology of logic. In this way, we would at least face a genuine and interesting choice between saying that the methodology of logic is more like that of the disciplines that count as "science" and saying that it is more like the methodology of the disciplines that don't count as "science." The question would be genuine because we would have concrete examples of disciplines with exceptional methodologies, and the question would be interesting because reasonable people could disagree on the question. Thus, I submit that the best way to formulate anti-exceptionalism is to specify some disciplines as having an exceptional methodology, where these disciplines should be such that a reasonable person could suggest that the methodology of logic is like that of these disciplines.

Two disciplines that can plausibly be taken to have an exceptional methodology and to be possibly similar to logic are mathematics and philosophy.[8] And logic is

[8] As an anonymous referee points out, philosophers like Ladyman and Ross (2007) might hold that insofar as philosophy (or at least metaphysics) has an exceptional methodology, this is a problem that must be overcome by changing the methodology of philosophy (or at least metaphysics). And one might hold that it is especially the *aprioricity* of philosophy's method that is problematic. I reject anti-exceptionalism about philosophy. However, this isn't relevant here because my thesis is that either one should accept global anti-exceptionalism, in which case debates about anti-exceptionalism about logic should be shaped by the correct global version of anti-exceptionalism and the arguments for it, or one should reject anti-exceptionalism about logic. My topic here is anti-exceptionalism that is specific to logic.

It may be worth mentioning in passing, however, that it seems to me that Ladyman and Ross (2007) run into a problem that is parallel to the problem that Martin and Hjortland are facing: there is no good philosophical account of what *the scientific method* is. Ladyman and Ross (2007, 28–38) end up appealing to peer-review processes and funding by established funding agencies to delineate what counts as science. While I recognize the authority of reason, as it is partially realized in science, I have little trust in peer-review processes and funding decisions; and I do not think that many scientists share the faith of Ladyman and Ross on these issues. If the question at hand is what reason demands of us when we do logic (or metaphysics), rather than what referees and funding

often taken to overlap with both disciplines (as illustrated in the quote from Shapiro above), thus making it reasonable for someone to hypothesize that logic is similar to mathematics or to philosophy in its methodology. So I suggest that the best way to formulate a genuine and interesting question in the vicinity of anti-exceptionalism is as follows:

> *Question*: Is the methodology of logic more like the methodology of mathematics and philosophy or more like the methodology of other disciplines, like the empirical sciences?

If anti-exceptionalists could establish that the correct answer to this question is that the methodology of logic is more like that of other disciplines than that of mathematics and philosophy, this would be an interesting and potentially illuminating result. Of course, we would still need to know in what way the methodology of logic is more similar to those other disciplines than to the methodology of mathematics and philosophy. Focusing as a first step, however, on the question above is still progress in the debate about anti-exceptionalism, or so it seems to me.

My thesis in this paper is that the methodology of logic is partially like that of mathematics and partially like that of philosophy. Since the methodologies of mathematics and philosophy are exceptional (as we assume in order to keep anti-exceptionalism interesting), it follows that the methodology of logic is exceptional. Hence, insofar as anti-exceptionalists deny that the methodology of logic is like that of mathematics or philosophy (as I just suggested they should), anti-exceptionalism is false. I will not try to rule out the possibility that logic has, besides a mathematical and a philosophical part, also a part that is similar to science in its methodology. Nor will I try to rule out that logic is like science in some respect. For all I say here, such a limited and modest version of anti-exceptionalism may be true; but it is not the kind of anti-exceptionalism that could justify sweeping claims like those in the quote from Hjortland above.

3 Mathematical Logic is Similar to Mathematics

In the previous section, I argued that anti-exceptionalism is best understood as implying that the methodology of logic is unlike that of mathematics or philosophy. In this section, I argue that one part of logic, namely mathematical logic, is similar to mathematics in its methodology. I cannot speak with authority about this part of logic, as

agencies demand, then anti-exceptionalism about logic (or metaphysics) must be established on the basis of an account of these demands of reason, rather than some hand-waving in the direction of the allegedly unquestionable rationality of (caricatures of) scientists. Whatever methodology science might use, its rational authority cannot come from the mere fact that it is accepted by referees and funding agencies. It must be possible to argue for the rationality of this methodology directly. So, we can cut out the middle man of appealing to science and ask directly what the arguments for some particular methodology in logic or metaphysics are.

I am neither a mathematician nor a philosopher of mathematics. Accordingly, what I say in this section is meant as a reminder of some uncontroversial observations.

A lot of the work that is done today under the heading of "logic" belongs to mathematics. Most papers published in the *Journal of Mathematical Logic*, for instance, are contributions to mathematics. Even many papers in the more philosophically oriented *Journal of Symbolic Logic* have titles like "Elimination of Imaginaries in Ordered Abelian Groups with Bounded Regular Rank" (Vicaria, 2023). And such research can be classified under subject 03 of the *Mathematics Subject Classification* (MSC), which is "Mathematical logic and foundations." It thus seems that one part of logic is a branch of—or similar to a branch of—mathematics. Publications in this part of logic seem to be assessed in a way that is similar to how contributions to other fields of mathematics are assessed. And the research reported in such publications seems similar to that in other fields of mathematics. It is thus reasonable to hypothesize that this part of logic is similar to mathematics in its methodology.

This hypothesis is supported by the following three observations: First, the most important methodology in mathematics and the mathematical part of logic seems to be the method of constructing theories—by laying down axioms, inference rules, definitions, and the like—and using mathematical proofs to establish results in and about these theories.

Second, if we set aside special issues like the search for new axioms in set theory, we don't need to choose between theories in pure mathematics. Rather, we can understand non-equivalent, consistent theories as investigating different mathematical objects. We need not choose, for example, between theories of groups in which group operations are commutative and theories in which group operations are not always commutative. There are simply abelian groups and nonabelian groups, as well as theories of both kinds of group. The same holds for logical theories understood as theories of pure mathematics. From the perspective of pure mathematics, there is no pressure to choose between classical, intuitionistic, paraconsistent, connexive, or any other logic. We can view them as theories of different kind of consequence relations and pursue all of them simultaneously without seeing any tension between them. Even committed anti-exceptionalists like Priest accept that much.

> As pure logics, no logic is a rival of any other. They are all perfectly good abstract theories. (Priest, 2003, 458)

Priest goes on to add that logic has a canonical application, namely our "practices of inferring" (Priest, 2003, 458) or deductive "reasoning" (Priest, 2014, 215), and that different logics are not equally good as applied logics for this canonical application. Here it suffices to note the uncontroversial nature of my claim that in (pure) mathematics and the (pure) mathematical part of logic non-equivalent, consistent theories are not viewed as rival theories.

Third, in mathematics and in the mathematical part of logic, there are recognized central results, that are established by well-known proofs, which students in the fields have to learn and which serve as paradigms of what counts as progress and as important results in the field. In logic these results include Gödel's completeness and incompleteness results, the undecidability of validity in classical first-order logic,

Gentzen's Cut-elimination theorem, the Post-completeness of classical propositional logic, and the like. A common way to approach a research question is to try to adjust techniques that are familiar from these paradigmatic results in such a way that they can be applied to the question at hand. This seems similar to how research is done in most fields of mathematics, and many of the paradigmatic results of logic have close connections to other fields of mathematics. Gödel's incompleteness results, for example, have far reaching implications for arithmetic and set theory, in a way that results of the empirical sciences do not seem to have implications for research in pure mathematics. Similarly, Hrushovski's proof of the Mordell-Lang Conjecture is an example of results from mathematical logic having implications for geometry (see Sagi, 2021).

I conclude that one part of logic, namely the mathematical part of logic, is similar in its methodology to mathematics. Since the methodology of mathematics (whatever it may be) should count as exceptional, as explained in the previous section, the methodology of the mathematical part of logic should count as exceptional. Indeed, it is easy to identify some respects in which this methodology is unlike that of the empirical sciences: (a) Non-equivalent, consistent theories are not seen as rival theories among which one must choose. (b) Results are established by mathematical proofs and not by experiments or other empirical studies. (c) The paradigmatic results in the mathematical parts of logic interact with mathematical research in the way in which mathematical results interact with mathematical research and not in the way in which results in the empirical sciences interact with mathematical research.

An opponent might object that the methodology of logic is unlike that of mathematics because in mathematics we must presuppose notions of implication and consistency to study and compare different mathematical theories, while logic investigates these notions of implication and consistency themselves. Moreover, we face the problem of theory choice with respect to the theory of implication and consistency that we use to study and compare mathematical theories. Thus, in logic non-equivalent theories of implication and consistency are rival theories, and their role in mathematics is not similar to the role of other mathematical theories. My response is twofold: First, it is common practice in logic to use classical logic (and set theory, etc.) in the metatheory to study and compare different logical theories. Thus, the situation in mathematics and in the mathematical part of logic seems to be the same. They both presuppose notions of implication and consistency in carrying out their investigations. Second, just as different notions of implication and consistency are sometimes advocated in mathematics, like in intuitionistic mathematics, so it is sometimes suggested that we should use some nonclassical logic in the metatheory in which we study and compare different logics. The reliance on a theory of implication and consistency in the metatheory in mathematics and the mathematical parts of logic may seem problematic from a philosophical perspective, but it is a part of how both of these fields are usually pursued today and, hence, no reason to deny that mathematical logic is like mathematics in its methodology.

My claim that mathematical logic is like mathematics in its methodology is not meant to be surprising. The claim is worth making, however, because it has recently been suggested that: "if we want to understand a field's research goals, methodological procedures, and epistemology, we are best placed looking at the activities and decisions of its practitioners" (Martin, 2024, Sect. 1). If we look at the activities and decisions of practitioners of the mathematical part of logic, we find a very successful practice with relatively well established methodological standards but in which theory choice plays no prominent role. If we look, however, at other parts of logic, it may be far less clear that we find any successful and progressive practice with a shared methodology. The practice we find may be no more successful or progressive or methodologically unified than many sub-fields of philosophy. In that case, the hope that we can make progress in the philosophy of logic by "looking at the activities and decisions" of the practitioners of the non-mathematical parts of logic may be as reasonable as the hope that we can make progress in the philosophy of ethics[9] by "looking at the activities and decisions" of philosophers who specialize in ethics. That is what I will suggest in the remainder of this paper.

4 Philosophical Logic is Similar to Philosophy

Besides a part that is a branch of (or similar to) mathematics, logic also has a part that is a branch of (or similar to) philosophy. This is the part of logic in which we find disagreements about questions like the following: Is classical or intuitionistic or relevance or connexive or some other logic the (or a) correct logic? Is there more than one correct logic? Is logic normative for thinking or reasoning? Do theories of logical consequence aim to represent something that is independent of our uses of concepts? How should we react to semantic and mathematical paradoxes? What is the best logic of conditionals? How do logic and probability theory hang together? How are logic and mathematics related? Is higher-order logic an acceptable and genuine part of logic? Etc. Such questions are answered by (some) logical theories; and these theories give rival answers, so that the question of how we ought to choose logical theories arises.

In this section, I will begin to argue that the philosophical part of logic is like philosophy in its methodology. This is a difficult project because the methodology of philosophy is unclear. Indeed, there does not seem to be any established methodology of philosophy. In light of this difficulty, I will assume that, whatever the methodology of philosophy is, it is characteristic of this methodology that philosophy has the following features. And I will argue that these features are shared by the philosophical part of logic.[10]

[9] The philosophy of ethics—that is, philosophical reflection on what we do in ethics—seems to be a part of ethics itself, just as the philosophy of any other part of philosophy belongs to that part of philosophy itself. Philosophy is reflective and reflexive in this way. I think that the same is true of the philosophical part of logic: The philosophy of logic is part of the philosophical part of logic itself. Of course, one can study parts of philosophy "from the outside," as a social or historical phenomenon. But to gain any philosophical understanding of a part of philosophy one must engage in that part of philosophy directly and not "from the outside."

[10] I agree with Tajer (2022) that we find different methodologies in the philosophical part of logic, as, for instance, the methodologies in debates about intuitionism, semantic paradoxes, and conditional

1. There are historical texts and traditions (of which some are older than, say, one hundred years) that are important for research in the field today, in such a way that current research engages (more or less consciously and explicitly) with historical texts and traditions (and not just textbook summaries of them).
2. Research projects in the field typically do not include any dedicated period (temporally distinct from periods in which hypotheses are formulated and periods in which data are analyzed) of rigorous and systematic data collection (like experiments, surveys, systematic observations, or quantitative analysis of corpora), after which previously formulated hypotheses are tested against the gathered data, such that there are antecedent and explicitly formulated standards for data quality and the methods for testing hypothesis in light of the data (such as established methods of inferential statistics, standards of experimental design like the randomization of control vs treatment groups, measures of the quality of measurements like Cronbach Alpha, and the like).
3. There are (or at least seem to be) perennial disagreements about foundational issues in the field, and there is no agreed upon methodology for (at least in principle) settling (or at least setting aside) these disagreements, so as to reach a widespread consensus.
4. The major milestones of progress in the field are not discoveries that are still viewed as correct (at least as approximations) but rather ways of reconceptualizing the whole field, so that the entire history of the field and research in it appear in a new light.

Philosophy has (or at least seems to have) these four features. Regarding the first, studying philosophy involves reading many centuries old books, and there are, for example, thriving research projects in ethics and metaphysics that understand themselves as neo-Aristotelian or Kantian or the like.

Regarding the second feature, probably every philosopher who has written an application for a major research grant knows that planning philosophical research is peculiar, compared to research in empirical fields. In philosophical projects, we typically do not find the usual division of research into a period of developing a theory, a period of designing empirical studies for testing the theory, periods in which data are gathered by conducting (some of) these studies, and periods in which the theory is assessed in light of the data. There are occasional exceptions (like projects in "experimental" philosophy) but those are rare.

Regarding the third feature, disagreements over realism regarding abstract objects, whether virtue can be taught, whether we can know anything and if so how, and the like are ongoing and have been part of philosophy for millennia. Although regularly some philosophers or schools of philosophy claim that they have found a method to settle such disagreements (as when Kant (KrV, B xviii) claims that his philosophy promises to metaphysics the "secure course of a science"), and sometimes this

logics are very different. However, I think the same is true of different debates in philosophy, and I suggest that the four features below are common to many debates in the philosophical part of logic and philosophy.

involves gathering data (as suggested by "experimental" philosophy), these suggestions always fail, at least in the sense that many philosophers remain unconvinced.

Finally, the texts that make up the canon of philosophy are usually not seen as records of philosophical discoveries that are generally accepted as correct (or good approximations). The groundbreaking texts of philosophy are rather suggestions for doing philosophy in a new way. In his *Meditations*, Descartes suggested taking human subjectivity as the starting point of first philosophy. In the *Critique of Pure Reason*, Kant suggested to pursue philosophy as an investigation into the conditions of the possibility of certain cognitions. And in the *Tractatus*, Wittgenstein suggested that philosophy is really concerned with the analysis of language. None of these suggestions are accepted by the majority of philosophers today. This contrasts, for example, with Galileo's discovery that the acceleration of falling bodies is independent of their mass, Kepler's laws of planetary motion, Newton's laws of motion, Maxwell's equations of electromagnetism, Einstein's general theory of relativity, and Heisenberg's matrix mechanics, which are contributions that are still accepted today as correct (or as good approximations)—although philosophers of science often claim that science is less cumulative than scientists think it is (Laudan, 1983; Kuhn, 1996).

In Kuhnian terminology,[11] we might say that philosophy is in a perennially pre-paradigmatic state or, perhaps, in a perennial crisis. A mature science, in which paradigms make periods of normal science possible, does not have the four features above. Regarding the first feature, Kuhn claims that scientific education makes no use of a "library of classics" (Kuhn, 1996, 167). Regarding the second, the scientific paradigms of normal science provide agreed upon methods that allow researchers to plan the systematic gathering of data and their systematic use in testing theories. Normal science does not have the third feature because, according to Kuhn, scientists turn to philosophical and foundational issues in their fields only during "periods of acknowledged crisis" (Kuhn, 1996, 88). And as for the fourth feature, scientists see the history of their field (wrongly, according to Kuhn) as a history of cumulative progress.

If the four features above are characteristic of the exceptional methodology of philosophy, we can now ask whether the philosophical part of logic has these four features. If anti-exceptionalism is correct and the methodology of logic is like the methodology of the mature sciences and unlike that of philosophy, the philosophical part of logic should not have these four features.

It may seem that the philosophical part of logic does not share these four features with philosophy because it may seem that there are paradigms in philosophical logic. It may seem, for instance, that an Aristotelian paradigm has dominated philosophical logic for centuries before a new paradigm was established by Frege, Russell, Hilbert, Tarski, Gödel and others. Alternatively, one might think that there existed several paradigms in the philosophical part of logic in the twentieth century.

[11] I use an admittedly old-fashioned Kuhnian framework here. This is in broad agreement with anti-exceptionalists's preferences (against the philosophy of science typical of logical empiricism or Popper). As far as I can see, not much hangs on this, and my points could be made in many other frameworks in the philosophy of science (like those suggested by Longino or Kitcher or Cartwright etc.).

It is certainly true that the work of Frege, Russell, Hilbert, Tarski, Gödel, and similar figures are milestones of progress in logic. However, they are primarily milestones of progress in the mathematical part of logic. The axiomatic systems and analysis of mathematical notions of Frege and Russell, the proof theory of Hilbert, as well as Tarski's model theory, and Gödel's incompleteness results are universally recognized achievements. But these are primarily contributions to the mathematical part of logic. The philosophical ideas of these figures are not part of the accepted methodological background of philosophical logic today or during the twentieth century. Frege's distinction between sense and reference, Russell's logicism, Hilbert's formalism, Tarski's theory of truth, or Gödel's rationalism have always been as controversial as seems to be the fate of philosophical ideas quite generally.

Of course, there are important and long standing traditions in the philosophical part of logic. These are, however, no more paradigms as the important and long standing traditions in philosophy. As long as we take philosophy to be exceptional, the existence of such traditions does not speak against the exceptional nature of logic.

I claim that the philosophical part of logic has the four features that are characteristic of the methodology of philosophy. In particular, I maintain that:

(A) Historical texts and traditions shape current research in the philosophical part of logic.
(B) Research projects in the philosophical part of logic typically do not include dedicated and planned periods of systematic and rigorous gathering of data, which are then analyzed and used to test theories by established methods.
(C) There are perennial disagreements without any agreed upon methodology for settling them in the philosophical part of logic.
(D) The major milestones of progress in the philosophical part of logic are suggestions for reconceptualizing the whole field rather than (apparently) cumulative discoveries.

In contrast to what we saw Martin and Hjortland claim about anti-exceptionalism, that the methodology of the philosophical part of logic has these four features does not imply that logic is maximally general, or *a priori*, or formal, or fundamental, or analytic, or necessary. Nor does it imply realism or anti-realism about logic. As far as I can see, the view that I am advocating in this paper is neutral on all of these points.

Why should we believe that the philosophical part of logic has these four features? The piece of evidence that is most convincing to me personally is my own experience of doing research in the philosophical part of logic. Together with Robert Brandom (Hlobil and Brandom, 2024), I have coauthored a book in which we pulled together the research we pursued over several years—in collaboration with other members of the Research On Logical Expressivism (ROLE) group. (Ad A) This research was shaped by historical texts from Aristotle to Gentzen. (Ad B) Our research did not include any dedicated periods of gathering data, nor did we test any theories in light of systematically acquired data by established methods. (Ad C) We address perennial questions about the nature of logic for which we do not know of any agreed upon method for settling them. (Ad D) Our aim was to offer a new way of looking at research in philosophical logic, and the parts where we added results to existing research in a cumulative fashion are the mathematical parts of our project.

To elaborate a little bit, we suggest a new logical theory, which can accommodate nonmonotonic and nontransitive material consequence relations; and we present this logical theory as our preferred and "to be chosen" theory. However, we do this against a very specific philosophical background that includes particular commitments about meaning, consequence, reason, the nature of logic, and discursive practices. Our research in philosophical logic cannot be separated from our related philosophical research, and our philosophical views on logic. Thus, insofar as I can speak with authority about my own research in philosophical logic, I have one clear data point that speaks for the claims (A)–(D) above.

I use my own research as an example here because my interest in anti-exceptionalism is driven by my interest in how I ought to conduct my own research in logic. If anti-exceptionalism does not apply to my work but only to the work of "real logicians" (as some anti-exceptionalists have suggested to me in conversation), this undermines most of my interest in it. And if anti-exceptionalism implies that I ought to change the way in which I conduct my research in logic by ignoring historical texts, by systematically gathering and analyzing data, by adopting a paradigm that allows me to ignore foundational questions, and by seeing the progress of logic as a series of cumulative discoveries, then this would change my research in ways that would make me lose all interest in it.

An opponent might object (a) that perhaps my research in philosophical logic is unusual or special or simply misguided. (b) Someone else might object that I am not a good judge of my own practice. (c) Yet another opponent might worry that anecdotal evidence about oneself should not carry any weight, otherwise my opponents could simply pit their own anti-exceptionalism friendly firsthand experience against mine, thus yielding a stalemate. In order to circumvent these worries and not focus on my own research in potentially problematic ways, in the next and final section, I want to sketch a disagreement in the philosophical part of logic and its historical roots. I will argue that this example supports my claims (A)–(D) above.

5 An Example: Disagreements About the Liar

In this section, I want to sketch three positions regarding the liar paradox that are currently defended in the philosophical part of logic. I will argue that the disagreement between these three positions supports my claims from the last section because: (Ad A) These positions are informed by historical texts and traditions. (Ad B) Research projects on these three positions do not include the systematic gathering of data that are then used to test theories by established methods. (Ad C) The disagreement between these positions is the latest manifestation of a very old disagreement for whose resolution there are no agreed upon methods. (Ad D) What supports these different positions on the liar paradox are not accumulated discoveries but rather specific ways to view the nature and role of logic.

There are an enormous number of positions and subtleties in current debates about the liar paradox. Among these positions we find three that can be summarized (in a simplified way) as follows:

Dialetheism: The liar paradox should be solved by accepting the contradictory conclusions of the paradoxical arguments. We should revise classical logic by rejecting the principle that contradictions imply everything because such alleged implications do not correspond to good reasoning, whereas all the steps in the paradoxical arguments are good reasoning (Priest, 2006).

Classicism: The liar paradox should be solved by saying that there are exceptions to the T-schema: "ϕ" is true if and only if ϕ. We should keep all of classical logic because it underlies (or partly constitutes, or is indispensable for) our best theories of what reality is like (Williamson, 2017).

Cut-Rejection: The liar paradox should be solved by keeping all classical entailments but also accept all instances of the T-schema; we should rather reject the Cut-rule for meta-inferences. The Cut-rule can fail because it can happen that a collection of assertions and denials does not violate certain coherence norms on discourse that consequence relations aim to codify (i.e. the collection is "in-bounds") but one can neither assert nor deny the liar sentence while keeping the collection of speech acts coherent (Ripley, 2013, 2015; Cobreros et al., 2013).

The advocates of these three positions do not merely disagree on what the correct solution to the liar paradox is; they also understand logic and (logical) consequence in different ways. It is common to distinguish (although sometimes in somewhat different ways) between, at least, an epistemic, a metaphysical, and a semantic understanding of logic (Hlobil, 2020; Shapiro, 2005, 2007). For my current purposes, I want to distinguish these conceptions of logic as follows[12]:

- *Epistemic conception of logic*: Logical theories provide means (or necessary parts of means) for assessing or guiding or improving acts or activities that aim at knowledge or understanding, such as reasoning, deducing, proving, explaining, extracting information from theories, maintaining consistency in one's beliefs, giving definitions, making distinctions, and the like.

[12] Notice that the distinction between these conceptions cuts across the distinction between model-theoretic and proof-theoretic conceptions of consequence. Many inferentialists have a semantic conception of logic and pursue it in a proof-theoretic way. And, for example, Priest has an epistemic conception of logic but often focuses on model theory.

- *Metaphysical conception of logic*: Logical truths describe the most general features of reality or the most fundamental or general structure of reality.

- *Semantic conception of logic*: Logical theories give an account of the meanings of some (or all) expressions or the content of concepts, which typically includes (or consists in) an account of which arguments are good in virtue of the meanings of logical expressions.

As is usual with broad families of philosophical views with illustrious pedigree, the differences between these conceptions are not sharp, so that some versions of them may not be mutually exclusive, and they are not exhaustive. Moreover, different aspects of the work of important figures in the field may suggest that they adopted different conceptions of logic. There are aspects in Frege's work, for example, that can justify attributing to him each of the three conceptions of logic just listed. And one might even hypothesize that good philosophical accounts of logic should accommodate the insights of all three (and possibly more) conceptions of logic. The situation seems broadly similar to normative ethics, where we find consequentialist, deontological, and virtue ethical theories, or the situation in metaphysical debates about whether universals exist *ante rem*, *in rebus*, or *in mente*. In all of these cases, it seems to be part of the philosophical challenge posed by these fields to find illuminating ways to combine insights from different approaches. I suggest that we understand the three conceptions of logic just listed in a broadly similar way.

The three positions regarding the liar paradox above are defended against the background of different conceptions of logic. Priest (2006, 2014, 2020) defends Dialetheism with an epistemic understanding of consequence in the background.[13] Williamson (2013b, 2017, 2024) defends Classicism with a metaphysical understanding of consequence in the background. And Ripley (2013, 2015) defends Cut-Rejection with a semantic understanding of consequence in the background.

In order to show that the disagreement between these three approaches to the liar paradox has the four features of philosophical methodology listed above, I now want to place each approach in a historical context, of which the advocates of the respective approach may be unaware and which they may not be willing to endorse explicitly but which is, I think, nevertheless informing these approaches. I will begin with the epistemic conception of logic and then turn to the metaphysical conception of logic, ending with the semantic conception. Of course, I cannot do justice to the history of these conceptions here. I am merely offering some pointers that support my claims about the methodology of the philosophical part of logic.

---

[13] In private conversation, Priest insists that he means the same by "consequence" as does Williamson, and that they agree that logic is about what follows from what and why. I suspect that this is due to two factors. First, Priest and Williamson both think that consequence has metaphysical and epistemic importance. Priest lets his metaphysics be informed by his logic, and Williamson thinks that consequence plays an important role in scientific theorizing. Second, it can seem that one knows what one means by "follows from" and that philosophical disagreements about how to understand consequence are disagreements about the best philosophical theory of whatever it is that we mean by "follows from." Neither of these two points undermines my claims in this paper, and I can accept both.

5.1 The Epistemic Conception of Logic

The Aristotelian tradition conceives of logic as a tool, which one has to acquire before pursuing any special science (Aristotle, *Meta.* 1005b2). Logic does not have a proper subject matter, as Aristotle says that neither "rhetoric nor dialectic is the scientific study of any one separate subject: both are faculties for providing arguments" (Aristotle, *Rhet.* I, 1256a30).

Aquinas follows but also transforms this tradition when he says that logic is not only an art by which we can, in our acts of reason, "proceed in an orderly and easy manner and without error" but adds that this art is also a science namely the "science of reason" (Aquinas, *Expositio libri Posteriorum Analyticorum*, preface). Buridan disagrees with this view and says that logic is not a science in a narrow sense because knowledge of logic is not a "speculative disposition" (*habitus speciulativus*) but a practical disposition (Buridan, *Quaestionum Libri Porphyry*, 126, l. 155–6). For the knowledge of the logic we use (*logica utens*) consists in an ability—namely the ability to construct arguments—and not in holding a theoretical belief. The logic that we teach students (*logica docens*), however, is a theory and so a science of argumentation (Buridan, *Quaestionum Libri Porphyry*, 127, l. 195–6).

Most philosophers of the early modern period, such as Descartes and Locke, accepted the idea that the purpose of logic is to help us to arrive at knowledge. However, they held that syllogistic logic does not work well for this purpose, and Bacon (*New Organon*, 35) concluded that "current logic [...] is not useful, it is positively harmful." Kant also held an epistemic conception of logic. However, he did not think that logic can serve as a tool to arrive at new knowledge, he rather held that logic sets a minimal standard for the correctness of thoughts (a *canon*). Logic merely tells us when a cognition meets a necessary condition to be knowledge, namely that it is "in agreement with itself" (Kant, AA IX, 14).[14]

Priest's conception of logic fits into this broad tradition because he holds that logic is concerned with regulating or assessing reasoning. Priest (2006, 40) holds that the subject investigated by logic is "constituted by the norms of correct reasoning." These norms of our inferential practice constitute our *logica utens*, according to Priest (2014, 219). Logical theories are theories of these norms. These theories are the *logica docens* (Priest, 2014, 216).

> [O]ne needs to distinguish between reasoning or, better, the structure of norms that govern valid/good reasoning, which is the object of study, and our logical theory, which tries to give a theoretical account of this phenomenon. (Priest, 2006, 207)

Priest, like Buridan, holds that the *logica docens* is similar to science. The point of this science is, however, the improvement or assessment of our activity of reasoning (Priest, 2014, 220). We should accept the logical theory that accurately captures the norms of good reasoning.

---

[14] There are also social or dialogical versions of the epistemic conception of logic (Marion, 2009; Dutilh Novaes, 2016).

Against this background of an epistemic conception of logic, it makes sense to reject the principle that contradictions entail everything or the principle that logical truths are entailed by everything. For, there are no good pieces of reasoning that lead from contradictions to arbitrary conclusions or from arbitrary premises to logical truths.

> [L]ogic [...] is supposed to provide an account of correct reasoning. When seen in this light the full force of these absurdities [e.g. that arbitrary premises entail any logical truth, according to classical logic] can be appreciated. Anyone who actually reasoned from an arbitrary premise to, e.g., the infinity of prime numbers, would not last long in an undergraduate mathematics course. (Priest, 1979, 279)

If the subject investigated by logic are the norms governing deductive reasoning, then logical theories that posit a norm according to which one can reason from contradictions to arbitrary conclusions or from arbitrary premises to logical truths are clearly mistaken. The norms that actually govern deductive reasoning do not sanction reasoning like that. And Priest shows how we can use this insight to deal with the liar paradox.

5.2 *The Metaphysical Conception of Logic*

A second conception of logic sees logic as a part of, or closely related to, metaphysics. This tradition can be traced to neo-Platonism. Already Plotinus explicitly denies that logic is merely a tool for epistemic purposes, as the peripatetic tradition claimed.

> For indeed dialectic should not be thought to be a tool of philosophy. For it is not concerned with bare theorems and rules, but it is concerned with real things and, in a way, has Beings as its material. (Plotinus, *Enneads* 1.3.5, 10–12)

Augustine continues this tradition in a Christian context and holds that the structure we find in logic reflects the divine order of reality. The validity of syllogisms is not a matter of convention or of what is conducive to our aims but is rather a matter of the "system of things."

> The validity of syllogisms is not something instituted by humans, but observed and recorded by them [...]. It is built into the permanent and divinely instituted system of things. (Augustine, 1995, 113)

In the ninth century, Eriugena agrees that logic is grounded in the nature of things rather than human thought or discourse. He holds that logic is concerned with the way in which reality is structured (into species and genera) and that this structure of reality "did not arise from human contrivances, but was first implanted in nature (*in natura rerum*)" (Eriugena, *Periphyseon*, 749 A).

After the middle ages, the metaphysical conception of logic surfaces again in German idealism. Hegel thinks that once we see the topic of logic clearly, which Hegel calls "thought", we can realize that it is "subjective and objective at once" and that the categories are the "essences of being" (Hegel, 1986, 240).

Heidegger continues this line of thought and calls for a new and philosophical logic, saying that the term “tool” reveals a “superficial-technical conception of logic” (Heidegger, 1978, 31–1). He draws on Leibniz to identify the principle of identity (which grounds the law of non-contradiction) and the law of sufficient reason (principle of ground) as the two basic general laws of thought. His perhaps central claim is that the principle of sufficient reason, as a metaphysical principle, is more basic than, and the ultimate foundation of, the principle of identity and, hence, of all of logic. He takes this to be a reversal of the traditional view that takes logic to be more fundamental than metaphysics.

> Our claim is that the first principle of logic is the principle of ground. But this claim is not just the reversal of the traditional order, but it is spoken out of the radicalization of logic toward metaphysics. The principle of ground is not a rule and norm for asserting; it is rather the first principle of logic as metaphysics. (Heidegger, 1978, 282)

Heidegger claims that logic should be understood as metaphysics (at least in its foundations). He holds that logic as the study of thought in general must ultimately turn into a study of what is shared by everything that we think about, that is, a study of being or the most general structure of the world.[15]

In the current debate, Williamson is following (unintentionally) in the footsteps of Hegel and Heidegger when he writes: “For quantified modal logic [...], one of its many roles is to supply a central structural core to theories of modal metaphysics. [...] Those theories are not about our language or thought [...], except incidentally, since they are about everything whatsoever” (Williamson, 2013b, x). He thinks that the usual meta-linguistic understanding of logical consequence and logical truth is misleading (Williamson, 2017, 329). Logic is not about sentences or arguments or proofs. Rather, logical laws are truths about the world, truths that capture the most general features of reality.

> [T]he law of excluded middle, $\forall P(P \vee \neg P)$, is just a very general structural law about the mostly non-linguistic world—a basic law of both logic and metaphysics, whatever Tarski would have said. [...] $\forall X \exists x(Xx \rightarrow \forall y Xy)$ is another very general structural law about the mostly non-linguistic world—another law of both logic and metaphysics. (Williamson, 2024, Sect. 1)

According to this view, logical theories are not primarily theories of logical consequence but rather theories about “very general structural laws about [...] the world,” which are expressed by logically true sentences. While logical truth might be a special

---

[15] It might seem that the metaphysical conception of logic is tied to a realist understanding of logic (see Tahko, 2021). However, there is a sense in which Hegel and Heidegger do not have a realist conception of logic. For them logic is not independent of the structure of the understanding of knowing subjects. These views are compatible because they hold that the structure of the world and the structure of the understanding of knowing subjects are essentially related to each other. A kind of anti-realism about logic that seems to me to be genuinely incompatible with the metaphysical conception of logic is what Commandeur (2024) calls “Non-representationalism about Logic,” which is the claim that “logics do not represent any extra-systematic phenomenon” (where I think Commandeur means something like “entities or facts” by “phenomenon” and not necessarily something that appears to a subject).

case of logical consequence from a proof-theoretic perspective, Williamson holds that "metaphysically, logical consequence is just a means to the end of logical truth" (Williamson, 2024, Sect. 1).

The principles governing the truth predicate are not very general or fundamental, according to this view. After all, the truth predicate applies only to entities of a very special kind, such as sentences or propositions. Compared to principles featuring the truth predicate, the laws of first-order (or second-order) logic are much more general, as they apply to every object and relation. From this perspective, it makes sense to give up the T-schema as holding in full generality rather than to revise logic. The T-schema is an approximation of a local and special structural law, and the lesson of the liar paradox is that it is merely an approximation that fails in some special cases.

5.3 *The Semantic Conception of Logic*

The semantic conception of logic sees logic as concerned with a special kind of subject matter, namely (certain) contents or meanings. This conception of logic can be traced back (at least) to the Stoa.

In opposition to the peripatetic view that logic is not a part of philosophy but merely an instrument of philosophy, the Stoa saw logic as one of philosophy's three parts: physics, ethics, and logic (Hülser, 1987, fragments 20–26). According to the Stoa, logic has a proper subject matter, and this subject matter are things that "belong to speech." Boethius describes the Stoic position thus: "logic treats of propositions and syllogisms and things like that, for which there cannot stand in things that don't belong to speech [...]" (Hülser, 1987, fragment 32a). Logic is part of the study of conceptual acts, in particular the contents of such acts, which they called "sayables" (Barnes, 1999, 66). According to the Stoics, sayables are abstract entities that exist independently of us, and arguments are compounds of such sayables (Barnes et al., 1999, 121).

Bobzien (2021) has argued that Frege's account of thoughts is influenced by this Stoic view. For Frege, logic provides a language with which we can express the abstract objects that are thoughts. And Frege is explicitly taking up Leibniz's idea of a *characteristica universalis*, in his *Begriffsschrift*, when he says that he was "striving after a *lingua characterica* in the first instance for mathematics, not a calculus restricted to pure logic" (Frege, 1979, 12). In particular, Frege takes logic to provide the "logical cement that will bind [... the nonlogical concepts as] building stones firmly together" (Frege, 1979, 13). Thus, Frege sees logic as (among other things) providing a language that can be enriched with nonlogical concepts to express the contents of any science in a particularly rigorous and perspicuous way. And to do so, logic must include a theory of what concepts and contents are. Thus, logic includes a theory of meanings and contents, which Frege spells out in terms of sense, reference, functions, and objects.

Brandom (1994, 108–11) takes up and develops this aspect of Frege's conception of logic. Brandom holds that logic provides an account of the meanings of logical vocabulary, which are special because they allow us to make explicit what follows from what and what is incompatible with what.

Bilateralists like Ripley (2015, 2017a) and Restall (2005, 2008) take up and develop Brandom's ideas. They give accounts of the meaning of expressions in terms of the conditions under which sentences in which they occur can be asserted or denied without violating certain coherence norms on discourse: the conditions under which such assertions or denials are or aren't "out-of-bounds." In particular, they give accounts of the meaning of logical vocabulary in this format, as when they suggest that it is a (partial) account of the meaning of negation to say that asserting a negation is out-of-bounds just in case denying the negatum is out-of-bounds and denying a negation is out-of-bounds just in case asserting the negatum is out-of-bounds. According to this version of bilateralism, logical consequence is the relation that holds between two sets of sentences if and only if it is out of bounds to assert everything in the first set and deny everything in the second set, and this holds in virtue of the meaning of the logical vocabulary that occurs in these sentences.[16] Bilateralists typically think that the relevant kind of incoherence is a matter of the norms that govern our discursive practices.

> What it is for a bunch of premises to entail a bunch of conclusions is that if you assert the premises and deny the conclusions, then you're out of bounds. [...]. The role this is playing is as a constraint on what kinds of things people can get away with in the conversational positions that they adopt. For example, [...] reflexivity [of consequence] is the claim that asserting and denying the same thing is out of bounds. [...] Not that you can't do it; go ahead. But what you've done clashes in some way. It's to be ruled out by some coherence-based norms on assertion and denial. (Ripley, 2015, 28)

Given such a conception of logic, it makes sense to give up the Cut-rule in light of the liar paradox (although only Ripley and not Restall advocates doing so). For according to this view, the Cut-rule says that if a collection of assertions and denials is in-bounds, then, for any sentence, one can either add an assertion or add a denial of the sentence to the collection and it remains in-bounds.[17] It seems plausible, however, that adding either an assertion or a denial of the liar sentence renders any collection of assertions and denials out-of-bounds, including collections of assertions and denials that are in-bounds. If that is correct, we have a counterexample to the Cut-rule. Hence, the idea goes, we should respond to the liar paradox by acknowledging that there are some sentences, such as the liar sentence, that one can neither assert nor deny without making one's collections of assertions and denials out-of-bounds, even if it was in-bounds before asserting or denying the liar sentence. The assumption, codified in the Cut-rule, that there are no such sentences simply turns out to be mistaken.

---

[16] The kind of incoherence in question is the incoherence we can find even among mere suppositions (so Moorean-Paradox-style incoherence does not count) (see Ripley, 2017b).

[17] If this isn't obvious to you, read the Cut-rule contrapositively: If $\Gamma \nvdash \Delta$, then, for any A, either $\Gamma \nvdash A, \Delta$ or $\Gamma, A \nvdash \Delta$.

5.4 *Lessons from the Case Study*

What does this disagreement tell us about my theses (A)–(D) above? It seems to me that it lends support to all four theses, namely as follows.

(ad A) The three positions on the liar paradox are shaped by ways of thinking about logic and logical consequence that are rooted in particular historical traditions. I have offered a sketch of the ancient roots of the three conceptions of logic that underlie the three views on the liar paradox, namely the peripatetic, the neo-Platonist, and the Stoic tradition. And while some of the advocates of the three approaches (especially Williamson) to the liar paradox do not place themselves in the traditions that I identified, it seems to me that such connections to long standing traditions is illuminating and potentially productive for current research on the liar paradox. Going back to the original sources can be illuminating in disagreements about the liar paradox and conceptions of logic in a way in which reading original texts is only rarely illuminating in the empirical sciences.

(ad B) None of the advocates of the three approaches to the liar paradox has engaged in any systematic gathering of data that could be used to test their theories. They sometimes point to uncontroversial facts about certain arguments being good, or how apparently different paradoxes are really similar, or what someone could reasonably say, or what sentences are meaningful, and the like. Such appeals are very different, however, from appeal to data in the sciences, where it makes sense to ask scientists for their "raw data," or to demand that they specify the way in which they will analyze their data before gathering the data (as this is increasingly institutionalized by requiring studies to be preregistered). Neither Priest, nor Williamson, nor Ripley use rigorous methods (like inferential statistics) for testing logical theories against systematically gathered data. And their research projects do not include assigned periods of data gathering, or hypothesis testing.

(ad C) The disagreement between the epistemic, metaphysical, and semantic conception of logic can be traced to antiquity, and there are no agreed upon methods for settling this disagreement. Insofar as these three conceptions of logic underlie the disagreement between the three approaches to the liar paradox, the latter disagreement is a reflection of a perennial disagreement regarding which we cannot reach a widespread consensus by applying agreed upon methods. There does not seem to be any "given" that could settle this disagreement.

(ad D) The three conceptions of logic that lie behind the three approaches to the liar paradox are not discoveries that are generally accepted as correct. They are rather suggestions for how to view the nature and role of logic, thus reconceptualizing the whole field of logic.

If there is a way to rationally choose between Classicism, Dialetheism, and Cut-rejection as logical theories regarding the liar paradox, it seems to me that this choice must be informed by an assessment of the three underlying conceptions of logic. Such an assessment, however, strikes me as a philosophical task that must be addressed by whatever methods philosophy can offer. The theory choice that we are facing with respect to the liar paradox seems to be the kind of theory choice that we face in philosophy. Perhaps philosophy cannot offer any agreed upon methods for choosing theories in such cases, and perhaps this seems to some like a disappointing result. Pessimism about philosophy is, however, not a good reason to stick one's head in the anti-exceptionalist sand and to pretend that the philosophical part of logic is not like philosophy but rather like science. Nor is pessimism about philosophy a good reason to indulge in the myth of the logically given.

6 Conclusion

I have argued that logic is unlike science—and thus exceptional—in the sense that the methodology of logic is unlike that of the empirical sciences but rather partly like the methodology of mathematics and partly like the methodology of philosophy. In the mathematical part of logic, no need to choose between theories arises. And in the philosophical part of logic, the rational way to choose between theories is similar to the rational way to choose between philosophical theories (if such a way exists).

Perhaps some think it is insulting to claim that logic, in its philosophical part, is not a mature and developed science. That would be a mistake. Many of the activities that deserve our deepest respect and admiration are more like philosophy than like science. It can hardly be insulting, for example, to hold that logic may be similar to the fine arts, ethics, or history in important respects, for example, in being a manifestation of our ability to reflect on ourselves. I suggest that we do not let our view of logic be clouded by a misplaced worship of science and rather embrace its philosophical nature with open arms.

Acknowledgements I would like to thank Nabeel Hamid, Olivia Sultanescu, Katharina Nieswandt, Robert Brandom, Rea Golan, Shuhei Shimamura, Ryan Simonelli, Dax Hamouth, and Ori Schnider for very helpful comments and discussion.

References

Andersen, F. J. (2023). Countering justification holism in the epistemology of logic: The argument from pre-theoretic universality. *Australasian Journal of Logic, 20*(3), 375–396.

Aquinas, T. (1949). *Commentary on the Posterior Analytics of Aristotle*. Magi Books.

Arenhart, J. R. B. (2022). Logical anti-exceptionalism meets the "logic-as-models" approach. *Theoria, 88*(6), 1211–1227.

Aristotle. (2014). *Complete works of Aristotle, volume 2: The revised oxford translation*. In J. Barnes (Ed.). Princeton University Press.

Augustine. (1995). *De doctrina christiana*. Clarendon Press.

Bacon, F. (2000). *The new organon*. Cambridge University Press.

Barnes, J. (1999). Logic and language: Introduction. In K. Algra, J. Barnes, J. Mansfeld, & M. Schofield (Eds.), *The Cambridge history of Hellenistic philosophy* (pp. 65–76). Cambridge: Cambridge University Press.

Barnes, J., Bobzien, S., & Mignucci, M. (1999). Logic. In K. Algra, J. Barnes, J. Mansfeld, & M. Schofield (Eds.), *The Cambridge history of Hellenistic philosophy* (pp. 77–176). Cambridge University Press.

Beall, J., & Restall, G. (2006). *Logical pluralism*. Oxford University Press.

Bobzien, S. (2021). Frege plagiarized the stoics. In F. Leigh (Ed.), *Themes in Plato, Aristotle, and Hellenistic philosophy, keeling lectures 2011–2018* (pp. 149–206). University of Chicago Press.

Brandom, R. B. (1994). *Making it explicit: Reasoning, representing, and discursive commitment*. Harvard University Press.

Burgess, J. P. (1992). Proofs about proofs: A defense of classical logic. Part i: The aims of classical logic. In M. Detlefsen (Ed.), *Proof, logic, and formalization* (pp. 1–23). Routledge.

Buridan, J. (1986). Jan Buridan, Kommentarz do Isagogi Porfiriusza. *Przeglad Tomistyczny, 2*, 111–195.

Carlson, M. (2022). Anti-exceptionalism and the justification of basic logical principles. *Synthese, 200*(3), 1–19.

Cobreros, P., Egré, P., Ripley, D., & van Rooij, R. (2013). Reaching transparent truth. *Mind, 122*(488), 841–866.

Commandeur, L. (2024). Logical instrumentalism and anti-exceptionalism about logic. *Erkenntnis*, 1–21.

Costa, N. D., & Arenhart, J. R. B. (2018). Full-blooded anti-exceptionalism about logic. *Australasian Journal of Logic, 15*(2), 362–380.

Dicher, B. (2020). Reflective equilibrium on the fringe. *Dialectica, 74*(2), 247–270.

dos Santos, C. F. (2021). Intuitions, theory choice and the ameliorative character of logical theories. *Synthese, 199*(5–6), 12199–12223.

Dutilh Novaes, C. (2016). Reductio ad absurdum from a dialogical perspective. *Philosophical Studies, 173*(10), 2605–2628.

Emery, N. (2023). *Naturalism beyond the limits of science: How scientific methodology can and should shape philosophical theorizing*. Oxford University Press.

Eriugena, J. S. (1987). *Periphyseon: The division of nature*. Dumbarton Oaks.

Ferrari, F., Martin, B., & Sforza, M. P. F. (2023). Anti-exceptionalism about logic: An overview. *Synthese, 201*(2), 1–9.

Finn, S. (2019). The adoption problem and anti-exceptionalism about logic. *The Australasian Journal of Logic, 16*(7), 231–249.

Frege, G. (1979). *Posthumous writings*. Oxford: Basil Blackwell.

Hegel, G. W. F. (1986). *Werke in 20 Bänden. Band 19: Vorlesungen über die Geschichte der Philosophie II*. Suhrkamp, Frankfurt am Main.

Heidegger, M. (1978). *Metaphysische Anfangsgründe der Logik im Ausgang von Leibniz*. Gesamtausgabe, Bd. 26: Abteilung 2, Vorlesungen 1923–1944. V. Klostermann, Frankfurt a.M.

Hjortland, O. T. (2017). Anti-exceptionalism about logic. *Philosophical Studies, 174*(3), 631–658.

Hjortland, O. T. (2019). What counts as evidence for a logical theory? *Australasian Journal of Logic, 16*(7), 250–282.

Hlobil, U., & Brandom, R. B. (2024). *Reasons for logic. Logic for reasons: Pragmatics, semantics, and conceptual roles*. Routledge.

Hlobil, U. (2020). Limits of abductivism about logic. *Philosophy and Phenomenological Research, 103*(2), 320–340.

Hülser, K. (1987). *Die Fragmente zur Dialektik der Stoiker: Neue Sammlung der Texte mit deutscher Übersetzung und Kommentaren* (Vol. 1). Frommann-Holzboog.

Kant, I. (1900ff.). *Gesammelte Schriften*. Bände I-XXII hg. v. d. Königlich Preussische Akademie der Wissenschaften, Berlin 1900ff., Band XXIII hg. v. d. Deutschen Akademie der Wissenschaften, Berlin 1956, ab Band XXIV hg. v. d. Akademie der Wissenschaften zu Göttingen, Berlin 1966ff.

Kant, I. (1966). *Kritik der reinen Vernunft*. Reclam.

Kneale, W., & Kneale, M. (1962). *The development of logic*. Clarendon Press.

Kripke, S. A. (2023). The question of logic. *Mind*.

Kuhn, T. S. (1996). *The structure of scientific revolutions* (3rd ed.). University of Chicago Press.
Ladyman, J., & Ross, D. (2007). *Every thing must go: Metaphysics naturalized*. Oxford University Press.
Laudan, L. (1983). The demise of the demarcation problem. In R. S. Cohen & L. Laudan (Eds.), *Physics, philosophy and psychoanalysis: essays in honor of Adolf Grünbaum* (pp. 111–127). D. Reidel.
Marion, M. (2009). Why play logical games? In O. Majer, A. Pietarinen, & T. Tulenheimo (Eds.), *Games: Unifying logic, language, and philosophy* (pp. 3–26). Springer.
Martin, B., & Hjortland, O. T. (2024). Evidence in logic. In M. Lasonen-Aarnio & C. M. Littlejohn (Eds.), *Routledge handbook of the philosophy of evidence*. Routledge.
Martin, B. (2020). Identifying logical evidence. *Synthese, 198*(10), 9069–9095.
Martin, B., et al. (2024). The practice-based approach. In E. Brendel (Ed.), *The oxford handbook of the philosophy of logic*. Oxford University Press.
Martin, B., & Hjortland, O. (2020). Logical predictivism. *Journal of Philosophical Logic, 50*(2), 285–318.
Martin, B., & Hjortland, O. T. (2022). Anti-exceptionalism about logic as tradition rejection. *Synthese, 200*(2), 1–33.
McSweeney, M. M. (2021). The cost of closure: Logical realism, anti-exceptionalism, and theoretical equivalence. *Synthese, 199*, 12795–12817.
Molick, S. (2021). Logical theory revision through data underdetermination: An anti-exceptionalist exercise. *Principia: An International Journal of Epistemology, 25*(1).
Peregrin, J., & Svoboda, V. (2021). Moderate anti-exceptionalism and earthborn logic. *Synthese, 199*(3–4), 8781–8806.
Plotinus. (2018). *The enneads*. Cambridge University Press.
Priest, G. (1979). Two dogmas of Quineanism. *The Philosophical Quarterly, 29*(117), 289–301.
Priest, G. (2016). Logical disputes and the a priori. *Principios: Revista de filosofia, 23*(40), 29–57.
Priest, G. (2003). On alternative geometries, arithmetics, and logics; a tribute to Łukasiewicz. *Studia Logica, 74*(3), 441–468.
Priest, G. (2006). *In contradiction: A study of the transconsistent*. Oxford University Press.
Priest, G. (2014). *Revising logic* (pp. 211–223). Cambridge University Press.
Priest, G. (2020). Logical abductivism and non-deductive inference. *Synthese, 199*(1–2), 3207–3217.
Quine, W. V. O. (1951). Two dogmas of empiricism. *Philosophical Review, 60*(1), 20–43.
Quine, W. V. O. (1964). *Two dogmas of empiricism*. Harvard University Press.
Read, S. (2019). Anti-exceptionalism about logic. *The Australasian Journal of Logic, 16*(7), 298–318.
Restall, G. (2005). Multiple conclusions. In P. Hájek, L. Valdés-Villanueva & D. Westerstaahl (Eds.), *Logic, Methodology and Philosophy of Science: Proceedings of the Twelfth International Congress* (pp. 189–205). KCL Publications.
Restall, G. (2008). Assertion, denial, commitment, entitlement, and incompatibility (and some consequence). *Studies in Logic, 1*(1), 26–36.
Ripley, E. (2017b). Uncut. Unpublished manuscript.
Ripley, E. (2013). Paradoxes and failures of Cut. *Australasian Journal of Philosophy,91*(1), 139–164.
Ripley, E. (2015). Anything goes. *Topoi,34*(1), 25–36.
Ripley, E. (2017). Bilateralism, coherence, warrant. In F. Moltmann & M. Textor (Eds.), *Act-based conceptions of propositional content* (pp. 307–324). Oxford University Press.
Rossberg, M., & Shapiro, S. (2021). Logic and science: Science and logic. *Synthese, 199*(3–4), 6429–6454.
Russell, G. (2008). *Truth in virtue of meaning: a defence of the analytic/synthetic distinction*. Oxford University Press.
Russell, G. (2019). Deviance and vice: Strength as a theoretical virtue in the epistemology of logic. *Philosophy and Phenomenological Research, 99*(3), 548–563.

Sagi, G. (2021). Logic as a methodological discipline. *Synthese, 199*(3–4), 9725–9749.
Shapiro, S. (2005). Logical consequence, proof theory, and model theory. In S. Shapiro (Ed.), *Oxford handbook of philosophy of mathematics and logic* (pp. 651–670). Oxford University Press.
Shapiro, S. (2007). Necessity, meaning, and rationality: The notion of logical consequence. In D. Jacquette (Ed.), *A companion to philosophical logic, chapter 14* (pp. 225–240). Wiley & Sons.
Tahko, T. E. (2021). A survey of logical realism. *Synthese, 198*(5), 4775–4790.
Tajer, D. (2022). Anti-exceptionalism and methodological pluralism in logic. *Synthese, 200*(3), 1–21.
Vicaria, M. (2023). Elimination of imaginaries in ordered abelian groups with bounded regular rank. *The Journal of Symbolic Logic, 88*(4), 1639–1654.
Williamson, T. (2007). *The Philosophy of Philosophy*. Wiley-Blackwell.
Williamson, T. (2013). Anti-exceptionalism about philosophy. *Croatian Journal of Philosophy, 13*(1), 1–3.
Williamson, T. (2013). *Modal Logic as Metaphysics*. Oxford University Press.
Williamson, T. (2017). Semantic paradoxes and abductive methodology. In B. Armour-Garb (Ed.), *Reflections on the liar* (pp. 325–346). Oxford University Press.
Williamson, T., et al. (2024). Is logic about validity? In E. Brendel (Ed.), *The oxford handbook of the philosophy of logic*. Oxford University Press.
Woods, J. (2023). A sketchy logical conventionalism. *Aristotelian Society, 97*(1), 29–46.
Woods, J. (2019). Against reflective equilibrium for logical theorizing. *The Australasian Journal of Logic, 16*(7), 319–341.
Woods, J. (2019). Logical partisanhood. *Philosophical Studies, 176*(5), 1203–1224.
Wyatt, N., & Payette, G. (2018). How do logics explain? *Australasian Journal of Philosophy, 96*(1), 157–167.

MIX
Papier aus verantwortungsvollen Quellen
Paper from responsible sources
FSC® C105338

If you have any concerns about our products,
you can contact us on
ProductSafety@springernature.com

In case Publisher is established outside the EU,
the EU authorized representative is:
Springer Nature Customer Service Center GmbH
Europaplatz 3, 69115 Heidelberg, Germany

Printed by Libri Plureos GmbH
in Hamburg, Germany